Cancer Etiology, Diagnosis and Treatments Series

CANCER BIOLOGY: AN UPDATED GLOBAL OVERVIEW

CANCER ETIOLOGY, DIAGNOSIS AND TREATMENTS SERIES

Cell Apoptosis and Cancer
Albina W. Taylor (Editor)
2007. ISBN: 1-60021-506-8

Chronic Lymphocytic Leukemia Research Focus
Chadi Nabhan (Editor)
2007. ISBN: 1-60021-526-2

Cervical Cancer Research Trends
Eleanor P. Bankes (Editor)
2007. ISBN: 1-60021-648-x

Lung Cancer in Women
Varetta N. Torres (Editor)
2008. ISBN: 1-60021-659-5

Cancer Research at the Leading Edge
Ignatius K. Martakis (Editor)
2008. ISBN: 1-60021-728-1

Chronic Lymphocytic Leukemia: New Research
Inès B. Moreau (Editor)
2008. ISBN: 978-1-60456-081-7

Cancer and Stem Cells
Thomas Dittmar and Kurt S. Zander (Editors)
2008. ISBN: 978-1-60456-478-5

Cancer Prevention Research Trends
Louis Braun and Maximilian Lange (Editor)
2008. ISBN: 978-1-60456-639-0

Clinical, Genetic and Molecular Precursor Features in Colorectal Neoplasia
Kjetil Søreide and Håvard Søiland (Editors)
2008. ISBN: 978-1-60456-714-4

Human Polyomaviruses: Molecular Mechanisms for Transformation and their Association with Cancers
Ugo Moens, Marijke Van Gheule and Mona Johannessen
2009. ISBN: 978-1-60692-812-7

Molecular Therapy of Breast Cancer: Classicism Meets Modernity
Marc Lacroix
2009. ISBN: 978-1-60741-593-0

Aromatase Inhibitors: Types, Mode of Action and Indications
Jean R. Lamonte (Editor)
2009. ISBN: 978-1-60741-711-8

Anticancer Drugs: Design, Delivery and Pharmacology
Peter Spencer and Walter Holt (Editors)
2009. ISBN: 978-1-60741-004-1

Lung Cancer in Women
Varetta N. Torres (Editor)
2008. ISBN: 978-1-60692-765-6
(Online book)

Cancer Biology: An Updated Global Overview
Tarek H. EL-Metwally
2009. ISBN: 978-1-60876-193-7

Cancer Etiology, Diagnosis and Treatments Series

CANCER BIOLOGY: AN UPDATED GLOBAL OVERVIEW

TAREK H. EL-METWALLY

Nova Science Publishers, Inc.
New York

Copyright © 2009 by Nova Science Publishers, Inc.

All rights reserved. No part of this book may be reproduced, stored in a retrieval system or transmitted in any form or by any means: electronic, electrostatic, magnetic, tape, mechanical photocopying, recording or otherwise without the written permission of the Publisher.

For permission to use material from this book please contact us:
Telephone 631-231-7269; Fax 631-231-8175
Web Site: http://www.novapublishers.com

NOTICE TO THE READER

The Publisher has taken reasonable care in the preparation of this book, but makes no expressed or implied warranty of any kind and assumes no responsibility for any errors or omissions. No liability is assumed for incidental or consequential damages in connection with or arising out of information contained in this book. The Publisher shall not be liable for any special, consequential, or exemplary damages resulting, in whole or in part, from the readers' use of, or reliance upon, this material. Any parts of this book based on government reports are so indicated and copyright is claimed for those parts to the extent applicable to compilations of such works.

Independent verification should be sought for any data, advice or recommendations contained in this book. In addition, no responsibility is assumed by the publisher for any injury and/or damage to persons or property arising from any methods, products, instructions, ideas or otherwise contained in this publication.

This publication is designed to provide accurate and authoritative information with regard to the subject matter covered herein. It is sold with the clear understanding that the Publisher is not engaged in rendering legal or any other professional services. If legal or any other expert assistance is required, the services of a competent person should be sought. FROM A DECLARATION OF PARTICIPANTS JOINTLY ADOPTED BY A COMMITTEE OF THE AMERICAN BAR ASSOCIATION AND A COMMITTEE OF PUBLISHERS.

Library of Congress Cataloging-in-Publication Data

El-Metwally, Tarek H.
  Cancer biology : an updated global overview / Tarek H. El-Metwally.
      p. ; cm.
  Includes bibliographical references and index.
  ISBN 978-1-60876-193-7 (softcover)
  1. Cancer. I. Title.
  [DNLM: 1. Neoplasms--metabolism. 2. Neoplasms--etiology. 3. Tumor Markers, Biological--metabolism. QZ 200 E48c 2009]
    RC254.5.E4 2009
    616.99'4--dc22   2009029570

Published by Nova Science Publishers, Inc. ✛ New York

Contents

Introduction		ix
Chapter 1	Cancer Definition and Classification	1
Chapter 2	Cellular Carcinogenesis is a Monoclonal Multistep Threshold Phenomenon	3
Chapter 3	The Genetic and Molecular Changes Leading to Cancer Are of Five Types	5
Chapter 4	Causes of Cancer	9
Chapter 5	Cellular Changes Associating Malignant Transformation (Dedifferentiation)	41
Chapter 6	Tumor Recurrence and Metastasis	43
Chapter 7	Tumor Markers	47
Chapter 8	Natural Inhibitors of the Multistage Carcinogenesis	55
Chapter 9	Biochemical Bases of Current Anticancer Treatments	57
Conclusion		61
References		63
Index		77

Introduction

Cancer cell should be envisioned as a cell that went out of the humoral and neural control due to loss of its physiological and biochemical differentiation as a integral unit of the human body. The loss of differentiation (dedifferentiation) includes the uncontrolled unnecessary cancer cell division. Understanding dedifferentiation necessitates an introduction to the normal control of the cellular differentiation. Cell differentiation is the specialization of the unspecialized embryonic cells into specific cell type to perform specific function. Normal cells are differentiated to one degree or another and generally differentiated cells behave in an expected manner, i.e., even zygote and embryonic cells that are totipotent (can differentiate into the entire differentiated cell types found in the adult organism) or pluri- to uni-potent stem cells (can differentiate into specific type(s) of cells) are considered at a certain differentiation level; See below.

Zygote $\Rightarrow$ totipotent $\Rightarrow$ pluripotent $\Rightarrow$ multipotent $\Rightarrow$ unipotent stem $\Rightarrow$ terminally differentiated cell

Differentiation is due to differential expression of tissue-specific genes that are under control of:

- Tissue-specific transcription factors and post-transcriptional regulation.
- Appropriate spatial cell-cell and cell-matrix interactions.
- Other epigenetic mechanisms. Differentiation involves the assembly of specialized forms of repressive chromatin including; hypermethylation, linker histones, polycomb proteins and methyl-CpG-binding proteins.

These structures compartmentalize chromatin into functional domains and maintain the stability of the differentiated state through successive cell divisions.

House-keeping vs. tissue-specific genes

The genes that are expressed in all tissues alike are called housekeeping genes. Although in most cells, the DNA sequence content of nuclei remains unchanged as development proceeds, the repertoire of genes that are expressed in a given cell type becomes limited (i.e., tissue-specific). It also becomes more difficult to reactivate genes that are silenced in that cell type. This limitation is now known to reflect the imposition of epigenetic regulatory mechanisms on genes. The molecular mechanisms necessary to stably repress genes are gradually established as embryogenesis and post-embryonic development proceed.

On terminal differentiation, some cells such as neurons may completely lose their ability to divide. Other cells may retain a degree or another of well-controlled proliferative ability just for renewal such as hepatocytes and mucosal epithelium. Other cells are ever dividing such as bone marrow stem cells. All are still classified as differentiated cells.

Once a normal differentiated cell acquires DNA mutation, it is obligated to stop dividing until repairing DNA; otherwise, it is instructed to induce its-own death by apoptosis that is the only fate of a normal terminally differentiated cell. This is very essential to protect the body from the potential hazards of this cell, e.g., transformation into cancer cell, teratogenic birth defects or autoimmune disease. This process requires functional antioncogenes and differentiation stabilizing mechanisms involving; retinoids, vitamin D, dietary flavonoids and butyric acid (product of fermentation of dietary fibers in the colon). Abnormalities affecting dietary supply, metabolism, catabolism and receptor function of such mechanisms could predispose to cancer.

Therefore, proliferation is not the pathognomonic difference between cancer and normal cell, but rather cancer cells lose differentiation to one degree or another. Cancer cell loses such differentiation characteristics (i.e., become dedifferentiated including inability to apoptosize) and revert into an embryonic behavior of uncontrolled unnecessary autonomous cell division. Therefore, cancer is graded according to the degree to which it retains the original differentiation characteristics into well-, moderately-, poorly- and un-differentiated grades that correlates more aggressive behavior, bad prognosis and acquisition of new mutations.

Relationship between apoptosis and differentiation and their dependence on normally functioning antioncogenes for the colon epithelium is depicted below. The epithelial lining of the colonic mucosa is renewed approximately every 6 days. Proliferating cells, localized to the lower two-thirds of the villus crypt, produce ~10^7 cells every hour; these cells continuously migrate along the crypt-luminal axis, undergoing maturation, differentiation, and, finally, apoptosis and sloughing into the colonic lumen. In the proliferating zone, unipotent stem cells proliferate rapidly and are liable for DNA damage causing a high rate of apoptosis with functional p53. In the post-mitotic terminally differentiated zone, cells are exposed to DNA damaging hostile lumen environment and undergoes apoptosis in presence of functional p53; See, Figure 1. Cancer transformation dedifferentiates the normal cell morphologically and biochemically (including conversion into anaerobic metabolism), immortalizes it, and stimulates its uncontrolled proliferation without apoptotic tendency.

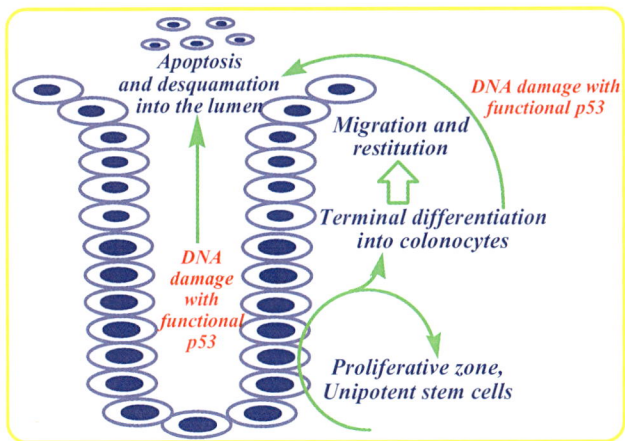

Figure 1. The normal fate of the colonic crypt stem cells and the terminally differentiated colonocytes in presence of functional tumor suppressor p53.

Chapter 1

Cancer Definition and Classification

Cancer or Malignant Tumors are composed of progressively dedifferentiating cells with a tendency of uncontrolled continuous, indefinite cell division and immortalization/ever-living as integral part of the dedifferentiation process. These pathological changes affect multicellular organisms, i.e., animals and plants. Cancer cells, termed *neoplastic (*neoplasm = neo/new and plasm/growth*), dedifferentiated* or *transformed cells*, do not perform useful functions in the body and do not obey the regulatory mechanisms controlling the rate of cell proliferation and differentiation. Cancer cell loses the collaborative integration of the multicellular organism and revert into autonomous unicellular organism behavior. Moreover, they could secrete effector molecules that disturb the body metabolism. Consequently, cancer cells form abnormal growth or tumor, which is potentially dangerous to the health and survival of the organism. Cancer is the second leading cause of death to heart disease.

Classification of Tumors

According to their embryonic tissue origin, tumors are either **carcinomas** (~80% of cancer incidence) originating from endo-/ecto-dermal tissues, e.g., skin and epithelial lining of internal organs and glands including cancers of the colon, breast, prostate, and lung; or **leukemias** and **lymphomas** (~9% of cancer incidence) the malignant tumors of hematopoietic cells of the bone marrow, and *Sarcomas* (~1% of cancer incidence) arising from mesodermal connective tissues, e.g., bone, fat, and cartilage. Except leukemia and lymphoma (liquid tumors), they all are called solid tumors because they grow in masses, whereas, leukemia proliferate as single cells. ***Benign tumors*** are tumors with good prognosis because their cells divide very slowly, are encapsulated in their place of origin and do not invade other layers. They keep most of the histological characteristic of the normal cells they originated from.

Malignant Tumors

Malignant tumors have bad prognosis because their cells divide faster and invade the surrounding layer to different degrees (tumor stages) and are able to distant metastasis by direct spreading and/or remote migration through blood or lymph. They lose the histological characteristic of the normal cells they originated from by variable degrees (grades) to complete loss, i.e., become undifferentiated cells. But among malignant cancers, there are massive variance in prognosis and so the most male attacking cancer is the prostate cancer and the most female attacking cancer is the breast one (33.5% each, of all cancers). However, this does *not* means that they are the most killing cancer, because as compared to lung (12.5%) and pancreatic cancer (2%), they are relatively of better prognosis since cancer pancreas kills within 1 year of diagnosis and lung cancer kills ~30% of all cancer deaths in both sex. The increased incidence of such hormone-dependent cancers (prostate, breast and genital) could be due to *increased* presence of tumor promoters (hormone and toxins with hormonal activity) in the human body *synthesized or diet-acquired*. Prognosis of malignant tumor then reflects the cocktail of mutation and mitogenic mechanisms involved in specific type of cancer.

Chapter 2

Cellular Carcinogenesis is a Monoclonal Multistep Threshold Phenomenon

Genetic instability and accumulation of a large number of mutations encountered in tumor cells reflect dysfunctional cell cycle regulating genes ("gatekeepers") and/or replication/recombination/mutation repair genes ("caretakers"). Therefore, cellular carcinogenesis is a complex process through stepwise progressive accumulation of genetic alterations involving oncogenes and tumor suppressor genes to a specific threshold. This is concurrent with progression through initiation, promotion, and progression (transformation, invasion and ultimately metastasis) with abnormal growth rate and developing drug resistance. The genetic processes leading to this progression disturb the equilibrium controlling the relative balance between inducers/inhibitors of; differentiation, the cell cycle, and apoptosis. The early stage of tumorigenesis is usually monoclonal, i.e., involving a single cell with the gain of oncogene activations and the functional loss of tumor suppressor genes by activating/inactivating mutation, deletion or protein-protein interaction; See, Figure 2.

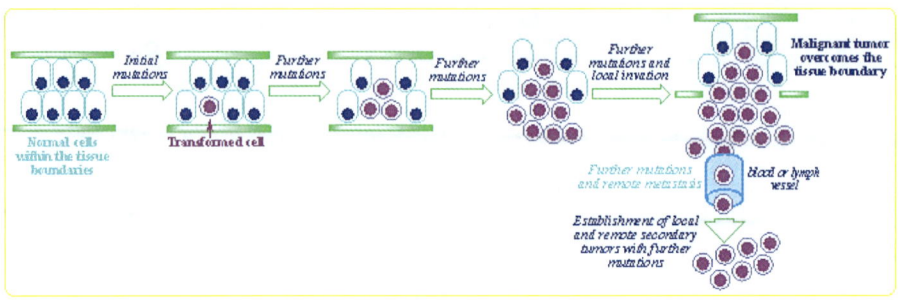

Figure 2. Progression from normal into local and remote secondary tumors requires sequential acquisition of mutations.

Mutations to cancer-associated genes may give cells a selective advantage - or make the cells resistant to die - which drives the evolution of sequentially more autonomous clones until a cancer has evolved. Although most malignant tumors are monoclonal in origin, their constituents are extremely heterogeneous. This is because multiple mutations occur independently as accumulating in different cells, thus producing subclones with different characteristics and a field effect in the affected tissue.

Chapter 3

The Genetic and Molecular Changes Leading to Cancer Are of Five Types

The genetic and molecular changes leading to cancer are of five types:

- **Dominant mutations** that activates normally functional and/or non-functional *proto-oncogenes* into abnormally functional and/or functional oncogenes, respectively. This type of activation could be due to; retroviral promoter and/or enhancer integration, point mutations, insertion mutations, gene amplification, chromosomal translocation and/or protein-protein interactions. Oncogenic gene products are active in stimulating cell proliferation, immortalization and/or dedifferentiation.
- **Recessive mutations** that inactivate normally functional anticancer, *tumor suppressor* or *anti-oncogenes*. Antioncogenic protein products are active at all levels of signal transduction cascades that inhibit cell proliferation and stimulate their redifferentiation and/or apoptosis.
- **Tumor-inducing viruses** are DNA containing, e.g., papilloma and adenoviruses or RNA containing (retroviruses), e.g., human T-cell leukemia viruses and the related retrovirus, human immunodeficiency virus. The retrovirus integrates its dcDNA form (created by its reverse transcriptase) into human cell genome with the help of their integrase. The retroviral strong promoter activity of their long terminal repeats (LTRs) flanking end of the transforming retrovirus transduces

downstream cellular proto-oncogenes into oncogenes. Viral oncogenes might had been acquired from human proto-oncogenes during the viral integration process, and vice versa. HIV induces certain forms of cancers by its integration-induced transformation. Cellular transformation by DNA tumor viruses is mostly due to protein-protein interaction; mainly by sequestration of tumor suppressor proteins by specific virus-encoded proteins (tumor antigens = T antigens) that inhibits the normal function of tumor suppressor proteins. A prominent example is the human papillomavirus E6 and E7 oncoproteins that sequester and hasten degradation of the antitumor p53 and Rb proteins. This virus is the prim cause of the cervical uterine carcinoma.

- ***Mutational inactivations of the DNA repair genes.*** Xeroderma Pigmentosum (XP) is an autosomal recessive disease due to *defective nucleotide excision repair* caused by mutation of at least 7 genes called *XPA – XPG* with helicase, kinase, recognition and excision activities. Patients are highly sensitive to sunlight and UV lights and their skin is damaged and is liable for multiple cancer, immunodeficiency, atrophy and scarring of eyelids, ulcerated cornea, and/or progressive neurological degeneration and may lead to premature death.
  o Ataxia Telangiectasia (AT) is caused by autosomal recessive mutations in the ATM gene (for Ataxia–Telangiectasia Mutated). ATM is a protein kinase known to be an "early responder" to double-stranded DNA breaks and its deficiency leads to increased sensitivity to X and other ionizing radiation-induced damage. In presence of such DNA damage, normal ATM protein (mainly acting through p53 protein) blocks progression through the cell cycle to induce repair or apoptosis. Cells of AT patients are resistant to apoptosis and do not undergo cell cycle arrest when subjected to ionizing radiation. Cerebellar degeneration, immunodeficiency, sterility, increased risk of cancer of the reticuloendothelial system and genomic instability are the characteristics of the disease.
  o Hereditary Non-polyposis Colorectal Cancer (HPCC) and Other Types of Cancer is also called Lynch syndrome in which patients develop cancer in the colon along with cancer of stomach, ovary and/or uterus. It is due to defective DNA mismatch repair with mutations in two genes, hMSH2 and hMLH1 (the human analogs of MutS and L). It has an autosomal dominant pattern of inheritance

with cancer at early age. Another mutation ascribed to this disease involves defective transcription-coupled nucleotide excision repair.
- Werner syndrome (WRN or Progeria of the adult) is an autosomal recessive syndrome caused by null mutations at a locus (WRN) coding for a member of the RecQ family of helicases. The gene product could play a role in such processes as DNA replication, DNA repair, DNA recombination, gene transcription and chromosome segregation. The presence of a 3'-5' exonuclease domain in the protein is consistent with a role in DNA repair. While WS is extremely rare (1-22 per million), the prevalence of heterozygous individuals may be as high as 1/150 to 1/200 individuals, at least in Japan. Somatic cells from such individuals have been shown to be hypersensitive to a powerful carcinogen and genotoxin, 4-nitroquinoline-1-oxide. The family members of WS pedigrees may be at higher risks for malignancy.
- Fanconi's anemia is an autosomal recessive anemia associated with increased susceptibility to chromosomal instability, pancytopenia and acute myeloid leukemia. It is due to defective correction of *chemical-induced DNA cross-links*. The FANC (A, B, C, D1, D2, E, F, G, I, J, and L - Fanconi's complementation groups) tumor suppressor genes are located to 11 complementation groups (e.g., FANCA at 16q24.3; FANCC at 9q22.3; FANCD2 at 3p25.3; FANCE at 11p15). They include components of DNA repair machinery where FANCD1 is the same for BRCA2.
- *Methylation of guanine* by alkylating agents is repaired by the "suicidal" O^6-*methylguanine-DNA methyltransferase* that transfers the methyl group from O^6-methylguanine into its own cysteine residue. This leads to irreversible inactivation of the enzyme. The gene of this enzyme is frequently repressed by hypermethylation in colon cancer. This permits the alkylating agents-induced GC-AT conversions. The later is implicated in the activation of the proto-oncogene K-ras into its oncogenic form seen in about half of colorectal carcinomas. This enzyme removes also chloroethyl and benzyl groups from guanine.
- The tumor-suppressor protein *BRCA1* (breast cancer susceptibility gene 1; often mutant in *breast and ovarian cancers*) is essential for Transcription-Coupled Repair associated with oxidative DNA

damage. The tumor-suppressor protein *BRCA1* has also been shown to play a critical role in non-homologous end-joining. The tumor suppressor protein *BRCA2* co-localizes with Rad51 during homologous recombinational repair, and contributes significantly to its activity.

- ***Metabolic derangement that mainly implicates the mitochondria at three levels.*** The first is the relative resistant to the induction of *mitochondrial membrane permeabilization*, which is the rate-limiting step of the intrinsic pathway of apoptosis. This is mainly due to the proapoptotic/antiapoptotic balance of the Bcl-2 family of proteins. This induces apoptosis resistance as one of the essential hallmarks of cancer. The second, cancer cell mitochondria often exhibit a *reduced oxidative phosphorylation*. This is reflected from the glycolytic mode of generation of ATP through the conversion of glucose to pyruvate and excess pyruvate is disposed as lactate even in conditions of high oxygen tension; i.e., aerobic anaerobic glycolysis. The third, cancer associated reduction in the mitochondrial efficiency also affects the *epigenetic pattern of the nuclear chromatin* that favors cancer poor prognosis through generation of more free radicals and lower energy.

Chapter 4

Causes of Cancer

Cancer is a multi-factorial-multi-stage disease that requires the cooperation of endogenous factors and the more important exogenous mutagens. Cancer may have a long period of latency that may reach 20 - 30 years from the time of exposure to mutagen till its clinical appearance. One in three persons will develop cancer in most developed countries where cancer is responsible for 25% of all deaths.

Endogenous factors include age, genetic susceptibility, immunologic, hormonal, endogenous mutagens, oxidative stress, angiogenesis and metabolic factors. *Exogenous mutagens* include; radiation, free radical generators, environmental and dietary chemical carcinogens and tumor promoters, oncogenic viruses, and certain bacteria and parasites. These are normally counteracted by anticancer factors, such as genetic resistance, DNA repair, antioncogenes, terminal differentiation, apoptosis, anti-tumor immunity and anticancer dietary nutrients. **Therefore, cancer is caused by the following environmental-genetic interactions:**

- Radiant energy (e.g. ionizing radiation, UV radiation, radon, and high frequency electromagnetic waves).
- Chemical carcinogens (e.g. pesticides, dioxin).
- Oncogenes (viral and cellular)/Tumor suppressor genes imbalance. And, certain bacterial and parasitic infections.
- Hereditary predisposition and age dependency.

Radiant Energy

Ultraviolet rays (UV), ionizing very short electromagnetic rays such as X-rays and γ-rays (and other electromagnetic rays) and even visible light in the presence of photosensitizers are mutagenic and hence carcinogenic in several ways. High energy particles such as α- and β-particles and neutrons are also ionizing with neutrons having great penetration power (up to a few mm) but α-particles penetrate soft tissues to a fraction of one mm.

- UV rays particularly the 280-320 nm range (UVB) generate oxygen free radicals and cause formation of pyrimidine dimers that lead to mutations if not properly repaired.
- X-rays cause non-genetic and genetic changes (DNA strand breaks), whereas, ionizing radiation such as γ-rays generate oxygen free radicals and subsequent DNA damage and mutation.
- The visible light, in presence of photosensitizers such as porphyrins and toxins, is able to produce mutagenic free radicals.
- High frequency electromagnetic waves may predispose to mutation and cancer.

Chemical Carcinogens and Carcinogenesis

Involvement of the environment in the occurrence of cancer has long been suspected as early as the 18th century, where high frequency of scrotum cancers was observed in chimney sweeps due to their professional environment. More recently, the role specific carcinogens play in cancer causation was established, e.g., tobacco and bronchiopulmonary cancer, asbestos and mesothelioma, UV radiation and melanoma, and, trichloroethylene and kidney cancer.

Chemical Carcinogens Are Classified on Several Bases

They could be complete or incomplete carcinogens; i.e., can induce a full blown cancer by itself or requires a concerted action with other carcinogens. They could be direct or indirect (procarcinogen) carcinogens; i.e., they act directly or they require preceding metabolic activation. They could be cancer initiators or promoter; i.e., they cause the initial immortalizing mutations or only promote the proliferation of such initiated cell. They could be classified according to their chemical structure or their known mode of action to transform the normal cells (direct genotoxic, indirect genotoxic and non-genotoxic). Both of these later two classifications are overlapping since more than one structural type could have the same mode of action and one structural type could have several modes of action.

Mechanism of Chemical Carcinogenesis: Initiation and Promotion

Carcinogens may enter the body in an active form, i.e., *direct carcinogen* such as ethidium bromide, acridines and nitrogen mustards (Figure 3). The nitrogen mustards, e.g., bis(2-chloroethyl)-ether and the (2-haloethyl) amines are cytotoxic DNA di- or poly-alkylating chemotherapy agents similar to mustard gas. Mustard gas alkylates guanine -NH_2 by their highly reactive sulfonium ion. The prototype nitrogen mustard drug is mustine which was the first drug to be used as an anticancer chemotherapeutic. Other nitrogen mustards include cyclophosphamide, chlorambucil, uramustine and melphalan.

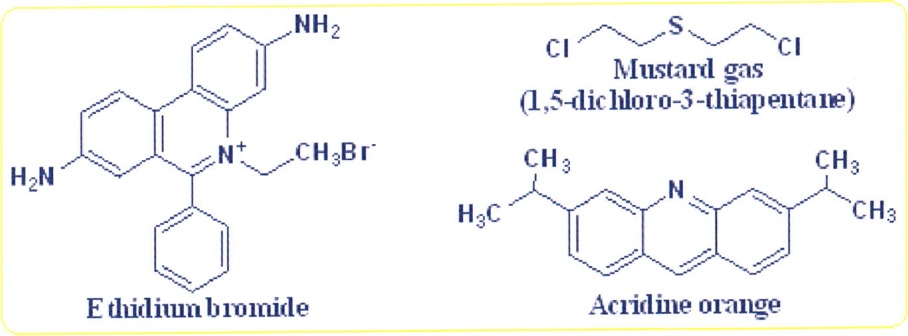

Figure 3. Formulae for some direct carcinogens.

Alternatively, a carcinogen may require metabolic activation into the active form from the inert form, i.e., *procarcinogen* such as the tobacco smoke carcinogen benzo(a)pyrene and the mycotoxin aflatoxin B1 (induce a very specific G-to-T mutation at codon 249 in the tumor suppressor gene p53). Procarcinogens require one or more steps of metabolic activation from the inert procarcinogenic form to the ultimate highly reactive carcinogenic form depending on the nature of the compound. Intermediate forms between the procarcinogenic and the ultimate carcinogenic form are called proximate carcinogen; See below for a hypothetical example.

Procarcinogen $\Rightarrow$ Proximate carcinogen A $\Rightarrow$ Proximate carcinogen B $\Rightarrow$ Ultimate carcinogen

Ultimate carcinogens are usually electrophiles, which covalently bind to the nucleophilic (electron rich) nitrogen and oxygen atoms in DNA, RNA and proteins causing adduct formation and damage to these macromolecules. The most dangerous result is the interaction with purines, pyrimidines or phosphodiesters of DNA. Their common site of attack is guanine. The establishment of mutations of different types including transitions, transversions, frame shift, deletion, insertion and point mutations, are the final mechanism of carcinogenesis. Examples are; dimethylnitrosamine that is activated in one oxidation step into a hydroxy reactive form that is able to alkylate (methylate) -OH or -NH$_2$ of guanosine; and, benzo(a)pyrene that is activated by several oxidation steps to the reactive diol-epoxide form that adducts onto NH$_2$-of guanosine; See Figure 4.

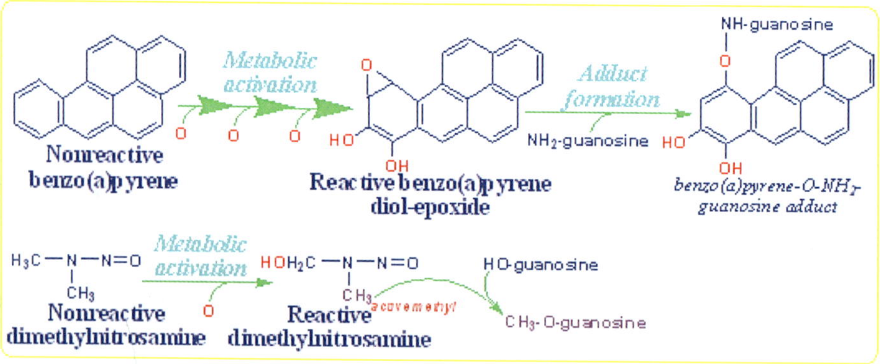

Figure 4. Activation of the procarcinogens; benzo(a)pyrene in several steps and dimethylnitrosamine in one step into their ultimate carcinogenic form.

Complete carcinogen, such as, asbestos and carcinogens in tobacco smoke, is able to induce the required cocktail of mutations to cause cancer in the lung without the help of other chemical or physiological promoters. Chemical or physical carcinogens are called *initiators* if they induce the first irreversible oncogenic mutation(s) of the cell that are not enough to induce full-scale malignant transformation. These initiators are consequently called *incomplete carcinogen*. Such initiated cells remain dormant until a favorable selective condition is available, where they proliferate and give rise to a tumor. Direct or procarcinogens could be complete or incomplete carcinogens. Therefore, carcinogens in general are genotoxic chemical (or physical or biological) agents that mutate DNA directly or indirectly through their active metabolite(s) or through other approaches, e.g., induction of genotoxic free radicals.

The physiologically active compounds that do not induce transformation but are able to promote a reversible stimulation of an initiated cell or microclone to grow into a full-scale tumor are called *promoters*. Examples of such promotion are the obligatory hormonal changes associating puberty of female rats that promote the carcinogenic action of dimethylbenzanthracene to cause mammary gland cancer in these rats. Other promoters include; phorbol esters (e.g., 12-O-Tetradecanoylphorbol-13-acetate (TPA; Figure 5), that mimics diacylglycerols in activating protein kinase C; PKC), croton oil (an essential oil rich in the tumor promoting phorbol esters), estrogens for breast and endometrial cancers and androgens for prostate cancer.

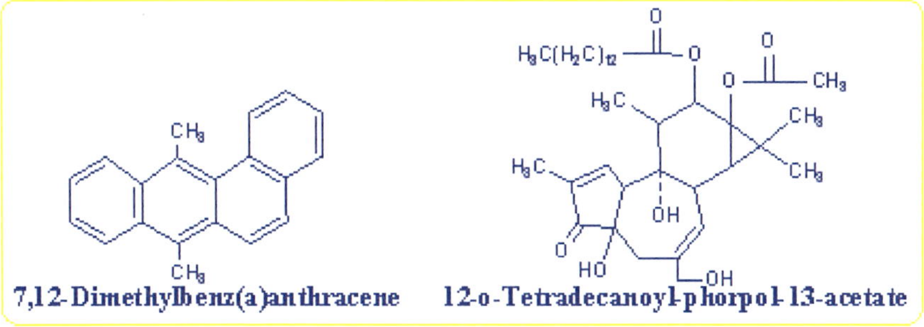

Figure 5. A cancer initiator - dimethylbenz(a)anthracene vs. a cancer promoter - phorbol acetate.

Thus, initiator acts to immortalize cells and promoter favors their proliferation. Application of an initiator, such as benzo(a)pyrene, to mice skin does not cause an evident tumor formation. However, the exposure of mice skin to benzo(a)pyrene followed by croton oil results in formation of malignant tumors. On the other hand, application of the promoter followed by the initiator results in formation of no or benign tumors of the skin, depending on the time lapse between the two events - due the reversible effect of the promoter.

Generally, promoters are non-genotoxic agents and include chemicals such as dioxin and organochlorine pesticides that work through legitimate/illegitimate cellular signaling pathway(s).

According to the pattern of exposure (ingestion, inhalation or transdermal absorption), chemical carcinogens are classified into:

1) Occupation hazards such as benzene and asbestos
2) Diet contaminants such as mycotoxins, formaldehyde, and pesticides (insecticides and herbicides)
3) Life style bad habits such as smoking, alcoholism and abnormal ways of cooking preserving foods; and
4) Drugs, e.g., base analogs (such as fluorouracil), alkylating agents (such as cyclophosphamide) and diethylstilbestrol.

According to the chemical nature, chemical carcinogens are classified into:

1) Polycyclic aromatic hydrocarbons, e.g., dimethylbenzanthracene and benzo(a)pyrene
2) Aromatic amines (dyes), e.g., 2-acetaminofluorene and N-methyl-4-aminoazobenzene
3) Nitrosamines, e.g., dimethylnitrosamine and diethylnitrosamine
4) Drugs, e.g., alkylating agents (such as cyclophosphamide) and diethylstilbestrol
5) Naturally occurring compounds and toxins, e.g., mycotoxins (e.g., aflatoxin B_1)
6) Inorganic compounds, e.g., arsenic, beryllium, cadmium and chromium salts
7) Non-alkylating agents, e.g., formaldehyde that cross-links DNA, RNA, and proteins; and nitrous acid that deaminates cytosine, adenine and guanine into uracil, hypoxanthine and xanthine
8) Alkylating agents, e.g., ethylmethane sulfonate that alkylates guanine
9) Fibers, e.g., asbestos
10) Mixtures: tobacco, environmental fine particles, and tars
11) Organochlorides and organobromides (pesticides, dioxin, polychlorinated biphenyls (PCBs), and polybromides); and
12) Organophosphorus compounds (sarin, and chlorpyrifos).

Some oncogenes and oncogenic viruses require chemical promoters such as mycotoxins, alcohol and smoking for hepatitis B to cause hepatocellular carcinoma; and, smoking, contraceptives, sunlight and genetic factors for papilloma viruses to cause ano-genital cancer. The impact of the different carcinogens, their sources and target tissues is listed in Table 1.

Table 1. The impact of the different carcinogens; Sources and target tissues

Group	Examples	Site of cancer
Environmental (5% of cancer deaths).	Aflatoxins. Erionite. Radon. Solar radiation (UV and visible).	Liver. Mesothelioma. Lung. Skin.
Life style habits & food additives (38% of cancer deaths)	Tobacco smoke (30% of cancer death). Alcoholic beverages (3% of cancer death) Dietary factors, e.g., artificial colors and flavors, salted fish, charring of meat, preservatives, low fiber, high animal fat and red meat contents (5% of cancer death).	Lung, larynx, mouth, esophagus, kidney, bladder, and pancreas. Head and neck, esophagus, and liver. Nasopharynx.
Hormones and reproductive (7% of cancer deaths)	Hormones (e.g., hormonal contraceptives, estrogens, androgens). Prenatal diethylstilbestrol.	Breast, endometrium, ovary, vagina, breast, prostate, testes, and liver. Vagina.

Table 1.(Continued)

Group	Examples	Site of cancer
Occupational (5% of cancer deaths)	Benzene.	Acute myelocytic leukemia.
	Benzidine.	Bladder.
	Arsenic.	Skin and lung.
	Chromium.	Lung.
	Asbestos.	Lung, pleura and peritoneum.
	Nickel dust.	Lung and nasal sinuses.
	Ionizing radiation.	Leukemia, lung, bladder, ovary, thyroid, bone, and soft tissues.
	Formaldehyde and Wood dust.	Nasopharynx.
	Aluminum manufacture.	Lung.
	Phenacetin.	Kidney and bladder.
	Sulfuric acid mist.	Nose and larynx.
	Nitrite.	Stomach.
	Aniline dye manufacturing.	Bladder.
	Nitrogen mustard gas.	Lung, head and neck, and nasal sinuses.
	Aromatic amines.	Bladder.
	Alkylating agents.	Bladder and leukemia.
	Polycyclic hydrocarbons.	Skin and lung.
	Vinyl chloride.	Liver angiosarcoma.
Drugs (2% of cancer deaths)	Alkylating agents.	Leukemia and bladder.
	Immunosuppressants.	Leukemia and lymphoma.
Infections (4% of cancer deaths)	Viruses (e.g., HPV, HIV, HCV, HBV, Epstein-Barr).	Urogenital, liver, skin, lymphoma.
	Bacteria (e.g., Helicobacter pylori).	Stomach.
	Parasites (e.g., Opisthorchus viverrini and schistosoma).	Cholangiocarcinoma and bladder.

Figure 6 depicts the sequential steps in the general carcinogen-mediated transformation (due to radiation, chemicals, inherited predisposition, mutations and chromosomal abnormalities or virally-induced) of a liver cell into cancer cell; illustrating the reversible and irreversible stages. Initiation and promotion changes and even early progression stages could be prevented and/or reversed by anticarcinogenic compounds, e.g., antioxidants, DNA repair and/or apoptosis of cells harboring mutation. Therefore, the conceptual 4 stages of the multistage carcinogenesis, namely; initiation, promotion, progression into malignant cell and metastasis - could be looked at as specific stage of accumulating mutations in the long path of reaching the required threshold to full-scale cancer. The accumulation of such mutations, and not necessarily the order in which they occur, constitutes multistage carcinogenesis. Permissive tumor microenvironment is essential for the establishment, local and remote invasion.

Figure 6. The reversible and irreversible steps in the general carcinogen-mediated transformation.

Classification of chemical carcinogens according to their major mode of action:

i) *Direct genotoxic* agents which cause DNA alterations directly and/or indirectly after metabolic activation. They include; physical agents, benzopyrene, and aflatoxin
ii) *Indirect genotoxic agents* which induce DNA alterations through cumulative or multiple types of damaging toxicities; they include tobacco particles, tar, asbestos and inflammation-inducers.
iii) *Non-genotoxic agents* which work through legitimate/illegitimate cellular signaling pathway(s). Examples of non-genotoxic agents include; dioxin and aromatic polycyclic hydrocarbons that utilize the xenobiotics aryl hydrocarbon receptor (AhR); pesticides that utilizes the pregnane X receptor (PXR); endocrine disturbing substances [e.g., estrogen mimicking substances and organochlorine pesticides such as dichlorodiphenyltrichloroethane (DDT) that activate estradiol receptor]; enzyme disturbing substances (organophosphorus compounds); cell stressors (oxidative stress, asbestos, metals, dioxin…). The effect of non-genotoxic agents culminates into uncontrolled cell proliferation and dysregulated metabolism. The normal function of such nuclear receptors (AhR and PXR receptors) is to adapt to xenobiotic influx by inducing detoxificating mechanisms.

Free Radicals and Cancer

One mechanism of inducing DNA alteration and cancer by certain carcinogens or non-carcinogenic chemicals is through inducing free radicals and oxidative stress. Reactive oxygen species are potential carcinogens because they facilitate mutagenesis, tumor promotion, and progression. Exposure to ionizing radiation has long been known to favor cancer development. Radiation-induced carcinogenesis appears to involve initiation, promotion and activation of protooncogenes and inactivation of differentiation stability and tumor-suppressor genes. Some genetic damage by radiation occurs by direct absorption of energy into DNA, but much is mediated by the formation of the highly reactive hydroxyl radical (OH$^•$). Hydroxyl radical is added into double bonds of DNA bases and abstracts H atom from the methyl group of thymine and each of the C-H bonds of 2'-deoxyribose. Thus, OH$^•$ could be involved in all stages of radiation-induced carcinogenesis. Several carcinogens are nongenotoxic (i.e., do not induce direct DNA alterations) even after metabolic activation, e.g., dioxin. However, they induce oxidative stress that in turn damages DNA and causes cancer.

Several mechanisms are utilized for this cause:

i) free radicals can be involved solely in a metabolic step eventually resulting in an ultimate chemical carcinogen
ii) the chemical itself or one of its metabolites can be converted into a radical that can react directly with DNA by an addition reaction
iii) the compound through its activating reactions or through "normal" metabolic pathways may cause the formation of free radicals which attack DNA; or, iv) following addition to DNA, the adduct can itself generate free radicals in proximity to the genetic material.

A role for free radicals and active states of oxygen in tumor promotion is supported by studies showing that:

i) free radical generating compounds act as promoters
ii) free radical scavengers act as promotion inhibitors, and
iii) endogenous cellular antioxidant defenses undergo changes correlating multi-steps carcinogenesis. *There are two basic ways that free radicals may be involved in tumor promotion:* i) by enhancing the expression of the somatic mutation or other growth-promoting processes laying between initiation and the development of a clinically recognizable tumor; or ii) by interference in the extracellular processes that normally inhibit cancer cell growth, for example by preferentially killing stromal non-cancer cells.

Although the role of free radicals in carcinogenesis is cleared out, it is still controversial, whether malignant tumors promote an oxidative stress *in vivo*. Tissue hypoxia, owing to an inadequate blood supply, is a common feature of most solid tumors. Differences in oxygenation within the tumor tissue may be one cause of this oxidative stress. These pO_2 differences range from 0.0 to >80.0 mmHg and may act like an intrinsic ischemia/reperfusion system. The ischemia/reperfusion condition, e.g., areas with changing oxygen pressure, is frequently present in tumor tissues. This condition is characterized by transient ischemia, resulting in ATP hydrolysis, hypoxanthine accumulation and transformation of xanthine dehydrogenase into xanthine oxidase. The following reperfusion with high oxygen pressure generates superoxide anion ($O_2^{\cdot-}$) and hydrogen peroxide (H_2O_2) due to oxidation of hypoxanthine by xanthine oxidase. Both, the presence of xanthine oxidase and elevated hypoxanthine concentrations have been described in tumor tissues. Due to the impaired energy metabolism in tumors, the accumulation of ADP and degradation of nucleic acids following cell death, promote the conditions for the production of significant amounts of superoxide anions during xanthine oxidase reaction.

The superoxide dismutation product, H_2O_2, is a well-established apoptotic agent and may therefore amplify the effect of xanthine oxidase. Furthermore, the reaction of reactive oxygen species with low molecular weight iron complexes would easily result in the onset of lipid peroxidation, and indeed the release of such iron complexes is likely to occur during disaggregation of dead cells. Thus, a series of biochemical changes set into motion within tumor tissue by either free radical formation or by boosting the ischemia/reperfusion condition would concur to favor the occurrence of oxidative cell injury and apoptosis.

Oncogenes/Tumor Suppressor Genes

Cancer-associated genes are; differentiation controlling (differentiation stabilizers/dedifferentiating), proliferation controlling (negative and positive effectors), or apoptosis controlling (pro-/anti-apoptotic). The first type of effectors are the tumor suppressor genes, whereas, the later type of effectors are the oncogenes. The main differences between oncogenes and tumor suppressor genes are presented in Table 2.

Table 2. Comparison between oncogenes and tumor suppressor genes (antioncogenes)

	Oncogenes	Antioncogenes
Nature of mutation:	- Dominant; one allele mutation.	- Recessive; two alleles inactivation.
Fate of mutation:	- Gain of function.	- Loss of function.
Location of mutation:	- Somatic cell, i.e., uninheritable.	- Somatic/Germ cell, i.e., inheritable.
Tissue preference:		
Genetic penetrance:	- Weak.	- Strong.

| Types:
Examples: | - Mostly weak with sporadic cancers.
- Cellular and oncoviruses.
- c- and v-ras, and, c- and v-erbB2 | - Mostly strong with familial cancers.
- Cellular.
- The p53 and Rb. |

Oncogenes

An oncogene is a gene - alone or in combination with other such genes - which is sufficient to inducing the cancer transformed phenotype of a normal cell. They are either of a cellular (designated *c-Oncogene*) or a viral (designated *v-Oncogene*) origin. However, the term "cellular oncogene" is a misnomer because these genes are not solely associated with cancer. In normal cells, proto-oncogene by being growth factors, members of signaling cascades of kinases, G-proteins, and other regulatory proteins that ultimately signal to transcription factors in the nucleus, regulate the expression of genes crucial in cell proliferation, cell differentiation, and cell death.

Most cellular oncogenes are composed of exons and introns, whereas, viral oncogenes look like processed genes without introns. This suggests that the virus might have acquired the oncogene through an intermediate mature RNA transcript. Nevertheless, the coding sequences of viral oncogenes and the corresponding proto-oncogenes exhibit a high degree of homology and very similar protein functions. Therefore, most probably, oncogenes (both viral and cellular) are derived from cellular genes. They encode various growth-controlling proteins active at all levels of the signal transduction cascades that stimulate dedifferentiation, immortalization and cell proliferation.

Proto-oncogenes were identified by being homologous to transforming retroviral oncogenes and by the tumorigenic ability of the transfection of their DNA into normal cells. The first evidence that oncogenes alone could induce malignant transformation came from studies of the v-*src* oncogene (encodes a 60-kDa protein kinase) from Rous sarcoma virus. When this oncogene was cloned and transfected into normal cells in culture, the cells underwent malignant transformation. Viral oncogenes are an integral part of the genome of cancer-causing DNA- and retro-viruses (oncogenic viruses). Retroviruses include human leukemia viruses 1 and 2, hepatitis C virus and human immunodeficiency viruses 1 and 2, and DNA-viruses include Herpes Epstein-Barr virus causing Burkitt's lymphoma; human papilloma viruses causing warts, skin and urogenital cancers; and hepatitis B virus causing hepatocellular carcinoma. These viral oncogenes are homologues of the c-oncogenes (e.g., v-/c-ras, v/c-erbB, and v-/c-erbA). Generally, viruses contribute to the etiology of only a few human tumors. Early vaccination against the most prevalent human papilloma virus types 16 and 18 greatly reduces the incidence of uterine cervical cancers.

In normal cells, oncogenes present in an inactive non-oncogenic form that is called *cellular proto-oncogene*. Qualitative or quantitative changes in the function of these genes - by mutations or defective transcriptional and posttranscriptional control - turn them into oncogenes. This leads to spatial, temporal, level and/or structural-functional changes in the target gene protein product.

Therefore, the abnormal oncogenic protein products of oncogenes are similar to substances normally involved in the control of cell division usually affect the different levels of a mitogenic growth factor signal transduction cascade. It could involve; amount of the available growth factors (e.g., PDGF); amount, affinity and structural alterations of the growth factor receptor (e.g., c-erbB2); amount and structural alterations affecting secondary transducing proteins (e.g., phosphatidylinositol-3-kinase and c-ras-G-protein); signal transducing non-receptor protein kinases (e.g., c-Raf); reaching finally to transcription factors that act directly to regulate the rate of gene expression (E2F, c-myc, and AP-1), and antiapototic effector (e.g, bcl-2).

Functional and Locational Classifications of Proto-Oncogenes

1. Growth Factors

The c-Sis gene (homolog to v-sis oncogene of simian sarcoma virus) located to 22q12.3–13.1 constitutively encodes the PDGF B chain that is reported mutant in glioma and fibrosarcoma. The c-Hst gene located to 11q13.3 constitutively encodes an FGF-related growth factor and was identified in gastric carcinoma. The int2 gene located to 11q13 constitutively encodes an FGF-related growth factor and was identified in mammary carcinoma.

2. Growth Factor Receptors

Gain-of-function mutations in the tyrosine kinase domain of the MET oncogene located to 7q31 leads to hereditary papillary renal carcinoma. It is the transmembrane receptor for hepatocyte growth factor (HGF)/scatter factor (SF). The MEN2A and B (Multiple Endocrine Neoplasia Type 2 or RET) protooncogene located to 10q11.2 was reported mutant in MEN2 - a dominant disorder characterized by pituitary adenomas, medullary carcinoma of the thyroid, and pheochromocytoma. It is the transmembrane tyrosine kinase receptor for glial-derived neurotrophic factor. The c-Mas gene located to 6q24–27 was reported in epidermoid mammary carcinoma encodes the G-protein-coupled angiotensin receptor. Some DNA rearrangement and/or point mutation result in fusion proteins/ligand-independent constitutive receptor activation. The c-Fms located to 5q33–34 encodes the colony stimulating factor-1 (CSF-1) receptor was reported constitutively activated by mutation in sarcoma. The c-Trk gene located to 1q32–41 encodes the NGF receptor-like protein and reported mutant (by DNA rearrangement leading to ligand-independent constitutive activation) in pancreatic, colon and thyroid cancers. The c-Kit gene located to 4q11–21 encodes the stem cell growth factor receptor was reported constitutively activated in sarcoma. The EGFR (c-erbB1), the epidermal growth factors located to 7p1.1–1.3 is broadly expressed in normal cells and its gene overexpression with increased protein level in many human cancers (including head and neck, lung, esophageal, gastric, pancreatic, colon, renal cell, breast, ovarian, cervical, and prostate cancers, and gliomas) correlates poor prognosis. The c-erbB2 (c-Neu/Her2) gene located to 17q11.2–12 encodes a member of the EGF receptor family proteins in ethylnitrosourea-induced neuroblastoma and breast carcinoma and several other cancer is activated by point mutation leading to a single amino acid change in the transmembrane domain and truncation (to become constitutionally ligand-independently active) or by gene amplification.

3. Membrane Associated Signal Transducing Non-Receptor Tyrosine Kinases

The v-/c-Src oncogenes located to 20p12–13 encodes a constitutively active protein tyrosine kinase and is reported mutant in colon carcinoma.

4. Signal Transducing Serine/Threonine Kinases

The c-Raf oncogene is functional with receptor tyrosine kinases, e.g., threonine phosphorylation and activation of MAP kinase.

5. Membrane Associated Signal Transducing G-Proteins

The homologs of the c-Ras oncogene (K-, the most important mutations, N-, the least important mutations, and H-, mutations are uncommon in human cancers) located to 1p11–13; 11p15.5 and 12p11.1–12.1, respectively, were identified mutant (mainly due to point mutation) in 20% of human cancers, e.g., colon, lung, pancreas, bladder and thyroid carcinomas, melanoma and myelogenous leukemia. They encode a constitutively active mutant GTP-bound G-protein with disrupted GTPase activity due to failure to respond to the regulatory interaction with GTPase activating proteins (GAPs) essential for exchanging GTP with GDP. They are most frequently reported mutant in colorectal and pancreatic carcinomas and melanoma.

6. Nuclear DNA-Binding/Transcription Factors

The c-myc oncogene located to 8q24.1 was first identified as homolog to the avian myelocytomatosis virus is involved in numerous hematopoietic neoplasias. Activation of its proto-oncogenic form is caused by retroviral integration and chromosomal rearrangements. N-myc form located to 2p24 and was reported amplified in neuroblastoma and lung carcinoma. L-myc form located to 1p32 was reported amplified in lung carcinoma. The AP-1 transcriptional regulatory complex reported mutant in sarcoma is a dimer formed of the c-fos oncogene (named after feline osteosarcoma virus) located to 14q21–22 and c-Jun oncogene located 1p31–32. The c-erbA1 and 2 located to 17p11–21 and 3p22–24.1 mutant forms of the thyroid hormone receptor were reported in erythroblastosis.

7. Antiapoptotic Factors

The bcl-2 oncogene (from B-cell follicular lymphoma and Epstein-Barr virus) located to 18q21.3 is involved in B-cell lymphoma and prevents initial apoptotic mechanisms at mitochondrial membrane level. It was mutated by chromosomal translocation to be under the control of the immunoglobulin gene enhancer.

Mechanisms of Activation of the Proto-Oncogenes

Mutations or genetic rearrangements of proto-oncogenes by carcinogens or viruses or epigenetically alter their normally regulated function into potent cancer-causing oncogenes. Such rearrangement includes six mechanisms. This process reflects the chromosomal instability of cancer cells and their defective DNA repair systems as major cancer-causing factors.

1. Viral Promoter/Enhancer Insertion

Viral integration into the host-cell genome may convert a proto-oncogene in its vicinity into a transforming oncogene. Example is the avian leukosis retrovirus, although it does not carry any viral oncogenes, yet is able to induce B-

`lymphomas through–integrating itself between exon 1 and 2 of c-myc proto-oncogene. On this position, it provides c-myc with a strong proviral promoter (i.e., the viral long terminal repeats, LTR). The c-myc activation in colorectal cancer and Burkitt's lymphoma caused by Epstein-Barr virus is another example. Also, the oncogenic DNA hepadna virus enhancer acting over a long distance might operate in liver cell transformation. Hepatitis B virus is involved in hepatocellular carcinoma and human papillomavirus is involved in cervical cancer. The human papillomavirus proteins; E6 and E7 bind and inactivate the cellular tumor suppressors; p53 and pRb, respectively.

2. Chromosomal Translocation

Ends of a double strand break in one chromosome are reciprocally translocated and joined to ends of a double strand break in the same or another chromosome by recombination. The simple forms of them are frequently seen in leukemia and lymphoma probably because these cell types normally rearrange their DNA to generate antibodies and antigen receptors. This process may bring forth an inactive proto-oncogene under the influence of a strong enhancer or promoter of a lost gene leading to altered function or expression. For example, c-myc is moved from its normal position on chromosome 8 to a position near the immunoglobulin heavy-chain enhancer on chromosome 14 in Burkitt's B cell lymphoma. The 15;17 translocation encodes a chimeric fusion protein, PML-RARα, by juxtaposing the retinoic acid receptor-α (RARα) gene from chromosome 17 and the promyelocytic leukemia (PML) gene from chromosome 15 PML with RARα in a head-to-tail configuration that is under the transcriptional control of PML. The protein binds to promoters containing retinoic acid response elements and recruits histone deacetylases (HDAC) to these promoters, effectively inhibiting gene expression rather than stimulating cell differentiation genes. This arrests differentiation at the promyelocyte stage and promotes tumor cell proliferation and survival. Treatment with pharmacologic doses of all-trans retinoic acid, the ligand for RARα, results in the release of HDAC activity and the recruitment of coactivators, which overcomes the differentiation block. The most important diagnostic marker of chronic myelogenous leukemia is identifying a clonal expansion of a hematopoietic stem cell possessing Philadelphia chromosome with unbalanced reciprocal t(9;22) that fuses head-to-tail the breakpoint cluster region (BCR) gene on chromosome 22q11 with the ABL (named after the Abelson's murine leukemia virus) gene located on chromosome 9q34. The chimeric gene is transcribed into a hybrid

BCR/ABL mRNA in which exon 1of ABL is replaced by variable numbers of 5'-BCR exons. Bcr/Abl fusion proteins, p210BCR/ABL contains NH_2-terminal domains of Bcr and the COOH-terminal domains of Abl. The abnormal fusion causes three critical functional changes; 1) the Abl protein becomes constitutively active as a tyrosine kinase that activates downstream kinases to prevent apoptosis and activates cell growth independent of external signals; 2) the DNA-protein-binding activity of Abl is attenuated; and, 3) the binding of Abl to cytoskeletal actin microfilaments is enhanced. Bcr/Abl fusion protein transforms hematopoietic progenitor cells *in vitro*. Also, in B-cell leukemia a translocation brings Bcl-2 under enhancer of immunoglobulin gene leading to the production of large amount of this oncogenic antiapoptotic factor.

3. Gene Amplification

Usually a gene has one copy per genome (but two alleles). Gene amplification means the increase in copy number of a gene per genome that occurs in 10% of late stages of tumors conferring bad prognosis. The amplified copies of the gene could stay in a linear tandem repeat at the chromosomal locus (causes homogeneous staining regions), arrange in onion-shell like pattern on top of the gene locus (forming a puff) or be in the form of an independent plasmid-like double minutes - double strand circular structures that are unstably presented in cancer cell and each contain two copies of the gene. Examples are the amplification of the receptors for certain growth factor, e.g., c-erb-B2 in brain glioma and breast tumors and N-myc in neuroblastoma. Tumor cells resist the action of the anticancer drug - methotrexate (an inhibitor of dihydrofolate reductase) by amplifying the enzyme gene into hundreds of copies. Increased level of enzyme production overcomes the competitive inhibition by the drug and acquire cancer cell an aggressive behavior and bad prognosis - along with other drug resistance strategies.

4. Point Mutation and Truncating Deletion

Mutation involving the regulatory and gene proper sequences is a major mechanism by which radiant and chemical carcinogens convert a proto-oncogene into a cancer-inducing oncogene, e.g., the missense point mutation at codons 12, 13 or 61 in the G-protein c-ras p-21 proto-oncogene in several human cancers. The mutation reduces the ability of the protein to associate with GTPase-stimulating proteins leading to its receptor-independent constitutional activation. Micro-satellite polymorphism instability also causes activation of c-ras. Nonsense point mutation and deletion that truncates certain subtypes of epidermal growth factor receptor (EGFR) - e.g., in brain glial tumors - leaves it without the extracellular ligand-binding domain and ligand-independently active.

5. Modulation of Activity of Cellular Receptors

Mutation-induced inactivation of antiproliferative growth controlling receptors such as TGFβ receptors leaves cells TGFβ-insensitive leading to uncontrolled proliferation as in colorectal and pancreatic cancers.

6. Altered Epigenetic Control of Gene Expression

Loss of genomic imprinting leads to increased gene dosage, e.g., for IGF-II in Wilm's renal tumors. Oppositely, repression of O^6-methylguanine-DNA methyl transferase gene - involved in the suicidal direct reversal of O^6-methylguanine DNA repair - by hypermethylation permits the activation of the proto-oncogene K−ras (by the alkylating agents-induced GC-to-AT conversions). This was reported in about half of colorectal carcinomas. Hippel–Lindau (*VHL*), breast cancer 1 (*BRCA1*), serine/threonine kinase 11 (*STK11*) genes and p16^{INK4a} are well-studied examples of tumor suppressor genes that are epigenetically silenced through hypermethylation in human cancers, e.g., renal, breast, and colon cancer.

Detailed Examples of Oncogenes

1. The Ras-p21 Protooncogene:

The G-protein ras/mitogen-activated protein (MAP) kinase pathway regulates multiple processes in cancer cells, including cell cycle progression, resistance to apoptotic signals, and cell motility. Mutations of the ras protooncogene occur in 20% of human cancers (e.g., cancers of pancreas, colon, and lung and myelogenous leukemia) and result in constitutively active ras with loss of its responsiveness to regulatory function of GTPase activating proteins (GAPs) to exchange GDP for the ras-bound GTP. The ras activates downstream effectors that include; ras/Raf serine-threonine kinase/MAP kinase and phosphatidylinositol-3-kinase (PI3K)/Akt pathways. The K-*ras* allele affected more commonly (85%) than N-*ras* (15%); H-*ras* mutations are uncommon in human cancers.

Ras activity in tumor cells is aberrantly increased by other mechanisms, including upregulation of the upstream receptor tyrosine kinase activity and mutation of GAP proteins, e.g., NF1 mutations in type 1 neurofibromatosis that act to inactivate ras. Epigenetic repression of O^6-methylguanine-DNA methyl transferase gene allows k-ras gene activation by alkylation mutation.

Ras proteins localize to the inner plasma membrane requires posttranslational modifications by farnesylation with farnesyl lipid moiety to the cysteine residue of the carboxy-terminal CAAX-box motif. Inhibition of farnesyltransferase prevents the membrane anchorage of ras particularly H-ras. However, lack of efficacy of these inhibitors appears to be due to ras alternative geranylation by covalent conjugation with geranylgeranyl molecule activated by geranylgeranyl transferase, which restores ras anchorage and function. Also, the mechanism of action of farnesyltransferase inhibitors appears to be due to inhibition of farnesylation of proteins other than ras, e.g., RhoB. Effector pathways downstream of ras include activation of the Raf serine/threonine kinase is induced by binding to ras and leads to activation of the MAP kinase pathway and cell proliferation.

2. The c-myc Protooncogene

The c-myc gene (after the original myeloid chicken leukemia retroviral oncogene) locates to 8q24 in human and is consisted of three exons with alternatively usable three promoters. Translation at the AUG start site in the second exon produces a major 439 amino acid, 64 kDa c-myc protein. Alternative translational initiation start sites result in both longer and shorter forms of the protein, termed p67 myc and mycS, respectively. The c-Myc protein is O-glycosylated and phosphorylated, and these modifications may alter the protein half-life. The c-Myc sequence contains several conserved N-terminal domains, termed myc boxes, which are found in closely related proteins, N-myc and L-myc.

The N-terminal transactivating domain is required both for transcriptional repression and induction; both are required in neoplastic transformation. The C-terminal helix-loop-helix leucine zipper (HLH-LZ) dimerization motif homodimerizes to another c-myc or heterodimerizes to other HLH-LZ proteins mainly max. Max is the obligate partner protein necessary for cellular transformation. For that c-myc competes for max with mad. Whereas c-myc-max dimer induces expression from E boxes (5'-CA[C/T]GTG-3')-containing genes, the mad-mad dimer mediates transcriptional silencing. For that both utilize histone acetylation/deacetylation. The short half-life of each of c-myc RNA and protein (30 min and 20 min, respectively) as compared to max and mad makes c-myc level to be the limiting factor and the regulated component of the heterodimers. Mad levels, as opposed to myc, increase during differentiation, and decreased expression of mad2 (mxi-1) protein was implicated in the development of cancer in a mouse model.

The expression of c-myc gene is tightly regulated in normal cells to be expressed only in actively dividing cells. The deregulated expression of c-myc due to genetic aberrations plays a significant role in human cancer development. The c-myc protein and/or gene is overexpressed in a wide variety of human cancers with 80% of breast cancers, 70% of colon cancer, 90% of gynecological cancers, 50% of hepatocellular carcinomas and a variety of hematological tumors possessing abnormal myc expression. This importance is reflected from the prognostic value of N-myc amplification in neuroblastoma, and of the chromosomal translocation affecting c-myc in Burkitt's lymphoma.

The resting cell expresses low level of c-myc, whereas, cells stimulated by mitogenic growth factors dramatically increase c-myc expression as an immediate early response gene. The high level persists throughout the cell cycle, but then returns to its basal quiescent state in resultant resting daughter cells. Abnormal or ectopic overexpression of c-myc in normal cells activates $p19/p14^{ARF}$ and a p53-dependent cell death pathway to guard against neoplastic transformation. Lack of both alleles of c-myc is incompatible with life causing early development death in mouse with a lack of primitive hematopoiesis and low proliferative ability of many cell types. For that, N-myc and c-myc are redundant, whereas, L-myc seemed dispensable.

There are several mechanisms for activation of the c-myc gene that contribute to the development of human cancers. *Chromosomal translocations*, as in the case of Burkitt's lymphoma, activate the c-myc locus by juxtaposing it adjacent to immunoglobulin genes that are transcriptionally highly active in B cells. *Gene amplification* increases myc gene copy number, which in turn increases expression of all members of the myc family. Over 200 copies per cell of N-myc can be found in neuroblastoma, and over 50 copies per cell of c-myc, N-myc, or L-myc may be found in small cell lung cancers. *Increased c-myc gene transcription* may account for the observed elevation of myc in human colon carcinoma. In colon, liver and other cancers, inactivating mutation of adenomatous polyposis coli (APC) and activating mutation of the transcriptional co-activator β-catenin, makes the later free to activate myc expression. *Removal of the destabilizing 3'-UTR sequences* stabilizes myc mRNA for higher translation. *Insertion of retroviruses* adjacent to the myc locus activates its expression via retroviral regulatory sequences. Oncogenic ras may stabilize the myc protein.

The c-myc has a role in cell cycle progression, metabolism, apoptosis and genomic instability. Acceleration of the cell cycle G1-S transition through abrogating cell cycle checkpoints and increasing cell metabolism by c-myc promotes cell proliferation and genomic instability (gene amplification, aneuploidy and polyploidy). For that, c-myc induces cyclins D and E, CDK4, and cdc25A. The later is a phosphatase that activates CDK2 and CDK4 and inactivates the cdkI p27. N-myc also activates telomerase that immortalizes neuroblastoma cells. *Normally this will be counteracted by induction of protective apoptosis pathways. However, in cancer cells with many additional mutations that are anti-apoptotic, c-myc can lead to full-blown neoplastic transformation.* Induction of the generation of reactive oxygen species from the mitochondria by c-myc is another cause of genomic instability due to damage to DNA. Metabolically, c-myc also induces the characteristic aerobic anaerobic glycolysis and synthesis of nucleotides in cancer cells.

Tumor Suppresser Genes (or, Anti-Oncogenes)

A tumor Suppresser gene (anti-oncogenes) is the gene that stabilizes cell differentiation and DNA integrity, prevents the progression into cell cycle for division and/or induces apoptosis to counteract the action of oncogene(s).

Inactivation of these antioncogenes is mainly due to point or large deletion mutation or epigenetic silencing by promoter hypermethylation. This results in uncontrolled cell proliferation of cells immortalized by proto-oncogene activation. Therefore, inherited malignancy predisposition and transforming activity of RNA/DNA tumor viruses and carcinogens may result from inactivation of anti-oncogenes and/or activating proto-oncogenes.

Functional and Locational Classifications of Tumor Suppressor Genes

1. Transmembrane Receptors

The PTCH (patched) tumor suppressor gene located to 9q22.3 was reported mutant in Gorlin Syndrome (Nevoid basal cell carcinoma syndrome) with basal cell skin cancer. It is the transmembrane receptor for sonic hedgehog (shh), involved in early development. DCC (Deleted in Colorectal Carcinoma) tumor suppressor gene is located to 18q21.3. It is a transmembrane receptor of the immunoglobulin cell-adhesion molecules superfamily involved in axonal guidance. It was reported mutant in colorectal cancer and pancreatic adenocarcinoma.

2. Regulators of Receptor-Associating Transducing Proteins

NF1 (neurofibromin 1) tumor suppressor gene is located to 17q11.2. It is a GAP that inactivates ras-G-protein and was reported mutant in neurofibromas, sarcomas and gliomas. The TSC2 (tuberin) tumor suppressor gene located to 16p13.3 was reported mutant in Tuberous Sclerosis 2 with benign growths (hamartomas) in many tissues, astrocytomas and rhabdomyosarcomas. It is a GTPase that activates RAP1 and RAB5. TSC1 (hamartin) tumor suppressor gene located to 9q34 was reported mutant in Tuberous Sclerosis 1 with facial angiofibromas. It interacts with tuberin.

3. Protein Kinase Inhibitors

The CDKN2A (cyclin-dependent kinase inhibitor 2A or p16^{INK4a}) tumor suppressor gene located to 9p21 was reported mutant in Familial Melanoma with pancreatic and other cancers. It inhibits cell-cycle kinases CDK4/6. The p57 (KIP2, kinase inhibitory protein 2 cell cycle regulator) tumor suppressor gene located to 11p15.5 was reported mutant in Beckwith-Wiedmann Syndrome with Wilms tumor, adrenocortical cancer and hepatoblastoma.

4. Protein Phosphatases/Kinases

The PTEN tumor suppressor gene located to 10q23.3 was reported mutant in Cowden syndrome with breast cancer, thyroid cancer, and head and neck squamous carcinomas. It is a phosphoinositide-3-phosphatase and a protein tyrosine phosphatase. STK11 (serine-tyrosine kinase 11 or LKB1 - a nuclear localized kinase) tumor suppressor gene is located to 19p13.3. It has a role in vascular endothelial growth factor (VEGF) signal transduction pathway. It also phosphorylates and activates AMP-activated kinase involved in stress responses, lipid and glucose metabolism. It was reported mutant in Peutz-Jeghers Syndrome with hyperpigmentation, multiple hamartomatous polyps, colorectal, breast and ovarian cancers.

5. DNA-Binding and Transcription Factors

P53 gene located to 17p13 is a cell cycle and apoptosis regulatory transcription factor. It was reported mutant in Li-Fraumeni syndrome, brain, lung, breast and colon tumors, sarcomas, leukemia, and most other types of cancer. RB1 transcriptional regulatory tumor suppressor gene located to 13q14 was reported mutant in familial retinoblastoma and most other human cancers including osteogenic sarcoma. WT1 transcriptional regulatory tumor suppressor gene located to 11p13 was reported mutant in pediatric Wilms' kidney tumors. DPC4 (Deleted in Pancreatic Carcinoma 4 and Familial juvenile polyposis syndrome; Smad4) tumor suppressor gene is located to 18q21.1. It is a downstream transcription factors for TGF-β/Bone Morphogenic Protein (BMP) signal transduction and was reported mutant in pancreatic carcinoma and colon cancer. Mutations affecting the receptors for TGF-β could have same cancerous consequences along with other developmental abnormalities as in colorectal cancer.

6. Proteins Involved in DNA Mutation Repair

BRCA1 breast cancer susceptibility tumor suppressor gene 1 located to 17q21 was reported mutant in familial breast and ovarian cancer. It regulates; cell cycle, protein degradation, transcription and DNA repair of double strand breaks by non-homologous end-joining. BRCA2 tumor suppressor gene located to 13q12.3 was reported mutant in familial breast and ovarian cancer. It regulates transcription of genes involved in DNA repair by homologous recombination of double strand breaks. The hMSH2 tumor suppressor gene located to 2p22-p21 was reported mutant in Hereditary Nonpolyposis Colorectal Cancer type 1 with DNA mismatch repair function. The hMLH1 tumor suppressor gene is located to 3p21.3. It has DNA mismatch repair function and was reported mutant in Hereditary Nonpolyposis Colorectal Cancer type 2. The ATM (Ataxia telangiectasia mutant) tumor suppressor gene is located to 11q22.3. It is involved in DNA mutation repair with p53 and was reported mutant in Ataxia telangiectasia with lymphoma, breast cancer, cerebellar ataxia, and immunodeficiency. The BLM (Bloom) tumor suppressor gene is located to 15q26.1. It has possible DNA helicase function important in DNA mutation repair. It was reported mutant in Bloom syndrome with solid tumors and immunodeficiency. The XP (Xeroderma Pigmentosum, A through G) tumor suppressor genes are located to 7 complementation groups (XPA at 9q22.3; XPC at 3p25; XPD at 19q13.2-q13.3; XPE at 11p12-p11; XPF at 16p13.3-p13.13).

They have DNA helicase activity functional in nucleotide excision repair and were reported mutant in Xeroderma Pigmentosum with skin cancer. The FANC (A, B, C, D1, D2, E, F, G, I, J, and L - Fanconi's complementation groups) tumor suppressor genes are located to 11 complementation groups (FANCA at 16q24.3; FANCC at 9q22.3; FANCD2 at 3p25.3; FANCE at 11p15). They include components of DNA repair machinery where FANCD1 is the same for BRCA2 and were reported mutant in Fanconi's anemia with chromosomal instability, pancytopenia, and acute myeloid leukemia.

7. Enhancers of Protein Degradarion

The VHL (von Hippel-Lindau) tumor suppressor gene located to 3p26-p25 was reported mutant in von Hippel-Lindau Syndrome with renal cancers, hemangioblastomas, pheochromocytoma and retinal angioma. It blocks transcription through VEGF/Hypoxia Induced Factors (HIFs) by enhancing their degradation thorugh activating ubiquitin ligase complex to enhance elongation of their polyubiquitinylation. This leads to high levels of HIF-1 and VEGF-induced angiogenesis with high microvascular density.

8. Proteins regulating mRNA Degradation

The HPC1 (or, PRCA1 RNaseL) tumor suppressor gene is located to 1q24-q25. It is involved in mRNA degradation and was reported mutant in hereditary prostate cancer.

9. Cell-Cell Adhesion and Signalling Proteins

The CDH1 (E-cadherin) tumor suppressor gene is located to 16q22.1. It is a cell-cell adhesion protein and was reported mutant in Familial diffuse-type gastric cancer with gastric cancer and lobular breast cancer. NF2 (neurofibromin 2) tumor suppressor gene is located to 22q12.2. It links cell membrane to cytoskeleton and was reported mutant in Schwann cell tumors, astrocytomas, meningiomas and ependynomas. APC (Adenomatous Polyposis Coli) tumor suppressor gene is located to 5q21. It is a signaling protein from focal adhesion receptors to the cytoskeleton and nucleus important for gaining differentiation and inducing apoptosis. Its germline mutation caused familial adenomatous polyposis type and somatic mutation causes sporadic type of colorectal cancer.

10. Proapoptotic Factors

The bax antioncogene located to 19q13.3-q13.4 is an example. Microsatellite instability that causes frameshift mutations in bax gene, deletion mutations and single nucleotide polymorphism mutations that affect the promoter of the bax gene significantly reduce its protein expression and activity and confers a growth advantage to the tumor cells by blocking the apoptosis program in several cancers including colorectal cancer, esophageal, lung, cancers and chronic lymphocytic leukemia and correlate disease progression and resistance to treatment.

Detailed Examples of Tumor Suppressor Genes

1. The p53 Gene

The p53, the "guardian of the genome," is a sequence-specific transcription factor whose activity is regulated through tight control of its protein levels. The nuclear localized phosphoprotein functions by binding as tetramer to 5'-PuPuC(A/T)(A/T)GPyPyPy-3' motif (where, Pu = purine and Py = pyrimidine). The level of p53 is low after mitosis but increases during G_1. During S phase the protein becomes phosphorylated by the M-phase cyclin-CDK complex of the cell cycle and also by casein kinase II (CKII). It transcriptionally regulates genes involved in suppression of cell growth, e.g., CDK inhibitory protein (CIP) - $p21^{CIP}$ that arrests G_1 and G_2 phase of the cell cycle. The involvement of p53 in both transcription and DNA replication and presence of various type mutations give rise to several probabilities ranging from mere recessive loss-of-the tumor suppressor function mutation to dominant gain-of-oncogenic function mutations.

Normally, levels of p53 are kept low by its association with the mdm2 oncogene product, which binds p53 and shuttles it out of the nucleus for proteolytic degradation. In response to ribonucleotide depletion, telomere shortening, hypoxia and DNA damage by, e.g., UV radiation, γ-irradiation or chemotherapy, p53 is phosphorylated by several kinases that regulate the DNA damage checkpoints. Phosphorylated p53 dissociates from mdm2, leading to increased intra-nuclear p53 protein levels that activates transcription of genes leading to cell cycle arrest ($p21^{Cip1/Waf1}$) or apoptosis (pro-apoptotic Bcl-2 family members, genes regulating metabolism of reactive oxygen species, and death receptors such as DR5). Further, deregulated activity of oncogenes such as *c-myc*, which promote aberrant G1/S transition, results in p53-induced apoptosis through $p14^{ARF}$, encoded by an *a*lternative *r*eading *f*rame from $p16^{Ink4a}$. The $p14^{ARF}$ is upregulated by *c-myc* and E2F, and binds to mdm2 to release p53 that induce cell apoptosis.

The mutant protein accumulates to high levels in tumor cells and often loses its DNA binding function. Incorporation of mutant protein in the tetrameric complex acts in a dominant manner to alter the whole complex's normal function. The p53 protein complexes with each of; SV40 large T antigen; E1B transforming protein of adenovirus; E6 protein of human papilloma viruses, and the oncogene mdm2 protein (located to 12q14, its gene is amplified) of the human sarcomas. This interaction prolongs the half-life of the wild-type protein a characteristic of its mutant forms and results in loss of p53 DNA binding and transcriptional activities. Acquired mutation in p53 is the most common genetic alteration found in human cancer (~50%) more frequent than any other known tumor suppressor or dominant proto-oncogene. It was demonstrated mutant in colon, breast, liver and lung cancers. Germline mutation in p53 is the causative genetic lesion of the Li-Fraumeni familial cancer syndrome. In many tumors, one p53 allele on chromosome 17p is deleted and the other is mutated. Inactivation of the p53 pathway compromises cell cycle arrest, attenuates apoptosis induced by DNA damage or other stimuli, and predisposes cells to chromosome instability. This genomic instability greatly increases the probability that p53 null cells will acquire additional mutations and become more malignant. Moreover, almost all human cancers have genetic alterations that bypass the Rb and p53 tumor-suppressor pathways.

Selenomethionine is the main dietary form of the cancer preventative selenium. It participates in a redox reaction resulting in the reduction of two cysteine residues within p53 that activate p53 DNA-binding to induce p53-depedent DNA repair pathways without affecting cell cycle.

2. Retinoblastoma (RB) Gene

The RB gene is composed of 27 exons (two are extremely large, 35 kb and 70 kb) that span 180 kb of chromosome 13. The 4.7 kb RB transcript encodes a p110 kDa protein (pRb) of 928 amino acids that is a nuclear localized phosphoprotein. The pRb is undetectable in retinoblastoma cells in which the two copies of the gene are mutant. Beside, the retinal cells, the mutation is also detectable in bone and connective tissue. The protein is detectable in most proliferating cells of other tissues of the retinoblastoma patient. Mutations include large scale deletions, splicing errors, point mutations and small deletions in the promoter region and mostly result in loss-of-function. In the familial form of retinoblastoma, the patient inherits a germline mutant allele mostly from the father and the disease development requires a subsequent somatic mutation to inactivate the remaining normal allele (a typical application of the double hit hypothesis).

The pRb regulates the cell cycle progression and is under regulation by inactivating phosphorylation mainly by G_1 cyclin-CDK complexes. Withdrawal of cell from cell division into differentiated state correlates hypophosphorylation of pRb but stimulation of quiescent cells with mitogen induces phosphorylation of pRb. The hypophosphorylated pRb complexes and inactivates the E2F family of transcription factors. Hyperphosphorylation of pRb by G_1 cyclin-CDK complexes upon mitogenic activation releases E2F to transcriptionally activate genes including S-phase cyclins and its own transcription. Transforming proteins produced during transformation by the DNA tumor SV40, adenoviruses, polyoma virus, human papillomavirus and BK virus bind and inactivate the hypophosphorylated pRb.

Hereditary Predisposition and Age-Dependency of Cancer

The cancer-causing genetic alterations are either vertically inherited as a germline mutation or may be horizontally acquired pre- or postnatal due to exposure to a carcinogen. The developing fetus is so sensitive to such mutagens due to its rapid cellular proliferation rate particularly in the first trimester.

Inherited germline mutation will affect all cells of the body, whereas, pre- or post-natal post-zygotic mutation will affect all or some of somatic cells. Examples of familial cancer syndromes include ~100 diseases with a high risk for malignancies and are mostly inherited in an autosomal dominant form but a few are autosomal recessive. They involve mainly tumor suppressor genes but a few affect oncogenes. They include; Xeroderma Pigmentosum, Ataxia Telangiectasia, Colorectal Cancer, Breast Cancer, Familial Polyposis Coli, Pediatric Lymphomas and Retinoblastoma. They occur early in age because they usually require one or a few more mutations to fully transform normal cells.

> ***The most significant risk factor for cancer overall is age*** with two-thirds of all cancer cases occurring over age 65. Cancer without hereditary predisposition (sporadic) usually occurs late in age (mid-seventies) because it requires accumulation of several mutations induced by environmental mutagens. Today, the high environmental pollution and unhealthy foods lowered the age threshold for sporadic cancers to be earlier (mid-fifties). For the interval 0-39 years of age, 1 in 72 men and 1 in 51 women will develop cancer; for the interval 40-59 years of age, 1 in 12 men and 1 in 11 women will develop cancer; and for the interval 60-79 years of age, 1 in 3 men and 1 in 5 women will develop cancer. But the rate seems to decline afterwards.

Familial cancers and pathologies that strongly predispose to cancer represent approximately ⸣5% of all cancers with inherited germline mutations. Although the genes causing this type of cancers have low prevalence, they confer high risk. Examples include; tumor suppressor genes (Rb/retinoblastoma; p53/Li-Fraumeni syndrome, and BRCA1 and 2/breast cancer; and, APC/colon cancer); and DNA repair enzymes (XPs/Xeroderma Pigmentosum; MLH1/Hereditary Non-Polyposis Colorectal Cancer; and, BLM helicase/Bloom's syndrome). Most of these are high penetrance monogenic cancer predisposing diseases.

Sporadic cancers, however, as the most common forms of cancer probably result from effects of multiple genetic variants of modest genetic impact but with high prevalence. They have uncharacterized inherited predisposition due to the weak penetrance of their genes and their polygenic nature with sporadic mutations in somatic and germline DNA. They include Single Nucleotide Polymorphisms (SNPs) that are genetic variants of a gene with a frequency of at least 1% in the general population). 20% of cancer cases can be attributed to weak-effect polymorphism with a high prevalence (50% of the population); that approaches the genetic impact of familial cancer-causing genes. They also include other risk factors, e.g., life style, occupational exposure, etc. The most studied SNPs are those involving genes regulating metabolism of xenobiotics (phase I enzymes such as cytp450 and phase II enzymes such as glutathione transferases and N-acetyltransferases). Other SNPs in other genes that can influence the risk of cancer include those involved in DNA repair, immunity, cell cycle control, or toxic substance dependency (addiction).

Examples of genes with causative SNPs changes include; xenobiotic phase I genes (e.g., cytp450s 1A1, 1A2, 2A6, 2D6, and 2E1; and alcohol dehydrogenases 2 and 3); xenobiotic phase II genes (e.g., glutathione transferases M1, T1, and P1; N-acetyltransferases 1 and 2; acetaldehyde dehydrogenase 2; sulfotransferase

1A1, and superoxide dismutase 2); DNA repair genes [e.g., X-ray repair cross-complementing group 1 gene (XRCC1) and 3; XPD and F; and excision repair cross-complementation group 1 gene (ERCC1)]; genes with role in immunity (e.g., IL-1α, 1β, 2, and 6; TNF; and HLA class I and II); cell cycle controlling genes (e.g., APC, p53 and ras-p21); and genes concerned with dependency to nicotine and other receptors [e.g., cytp450s-2A6; D2 dopamine receptor gene (DAT1), dopamine transporter 1 and 4 genes (DRD2 and 4); and the retinoids receptor RARα]. A well-documented example of this is the risk of bladder cancer and exposure to aromatic amines as a function of N-acetyltransferase 2 genotype (fast vs. slow acetylators) reflected as differences in cancer risk. Other genetic polymorphisms that modify the risk of cancer associated with environmental or occupational exposure include; glutathione transferase M1 in aflatoxin-induced liver cancer; acetaldehyde dehydrogenase 2 in alcohol-induced esophageal cancer; N-acetyltransferases 1 and 2 and sulfotransferase A1 in heterocyclic amines-induced colon and breast cancers; cytp450-A1 and 2, glutathione transferase M1, N-acetyltransferases 1 and 2, microsomal epoxide hydrolase gene (EPHX1; involved in both the activation and detoxification of several tobacco carcinogens) and XRCC1 in tobacco-induced lung and bladder cancers; and XPD in UV-induced skin cancer.

The likelihood of a first degree female relative of a woman with breast cancer to developing the disease is between 1.8 - 3 times the general population risks - being greater the younger the age of onset of the proband. Women who have had children have a lower risk of developing breast cancer than nulliparous women. And, the younger the age at which woman has her first pregnancy, the lower the risk of developing breast cancer. However, concordance rate for breast cancer, as an example, in both types of twins (mono- and di-zygotic) is about 15% suggesting that the environmental factors are likely to be more important than genetic factors.

Thus, 50 - 80% of human cancers are preventable. Primary prevention involves identifying and manipulating the causal genetic, biologic, and environmental factors. Essential are; smoking cessation, diet modification, antioncogenic viral vaccines, eradication of schistosomiasis and *Helicobacter pylori*, and chemoprevention (using, e.g., plant fibers, retinoids and flavonoids; antioxidants, and non-steroidal anti-inflammatory drugs, e.g., aspirin). Secondary prevention includes; screening of populations at risk (e.g., mammography and colonoscopy), identifying asymptomatic neoplastic lesions and their proper therapy, and prophylactic mastectomy in high risk women (genetically positive for BRCA1 and 2 mutation and familial history), and colonectomy in patients with familial polyposis and ulcerative colitis (genetically positive for APC and the HNPCC mismatch repair genes MSH2 and MLH1 mutations). Genetic testing for various mutant genes associated with the predisposition to cancer in affected families and in some subpopulations with a known increased risk, even without a defined family history, e.g., the two 185delAG and 5382insC mutations in the breast cancer susceptibility gene BRCA1 that exhibit a sufficiently high frequency in the Ashkenazi Jewish population. However, for many cancer predisposition genes, the sensitivity of genetic testing is only 70% or less because it could be non-informative for purely technical reasons and/or a mutation that could be aborted later through apoptosis.

Chapter 5

Cellular Changes Associating Malignant Transformation (Dedifferentiation)

Cancer cell is immortalized by activation of proto-oncogenes and is allowed to progress through cells cycle due to inactivation of anti-oncogenes. The increased expression of growth factors and/or their receptors give them the driving force for uncontrolled proliferation. To metastasize, cancer cells require further mutational changes such as structural membrane (affecting glycolipids and glycoproteins) and cytoskeletal alterations, and secretion of several proteases (See, Figure 7). The following are the transformation-associating changes.

1. Morphological dedifferentiation including; loss of polarity and increased nucleus to cytoplasm ratio and heterochromatin to euchromatin ratio). This is due to alteration of cytoskeletal (e.g., actins and cytokeratins) and membrane structures.
2. Loss of density-dependent (contact) inhibition of growth.
3. Loss of anchorage dependency (i.e., gain of the ability to growth in soft agar) due to altered cell-cell and cell-matrix communication. Transformed cells acquire mobility for local invasion and distance metastasis, i.e., resembling an autonomous unicellular organism.
4. Biochemical dedifferentiation including; increased rate of aerobic anaerobic glycolysis and reduced mitochondrial activity and/or function, secretion of certain proteases, alteration in structure of cell surface

antigens, alteration in isoenzyme profiles towards fetal pattern, synthesis of fetal proteins and reduced differentiation functions (i.e., dedifferentiation and diminished synthesis of specialized proteins, e.g., albumin synthesis for hepatocytes and carbonic anhydrase II activity in ductal pancreatic cells).
5. Diminished requirement for exogenous growth factors due to increased synthesis and secretion of growth factors and up-regulation of the growth factor receptors. This gives cancer cells one of their most important characteristics, i.e., autonomous growth (independence).
6. Potential immortalization, i.e., ever living and inhibition of apoptosis with ability to subculture cancer cells *in vitro* indefinitely.
7. Disordered pattern of growth and mitosis and abnormal chromosomal number (aneuploidy) and structure.
8. Tumorigenicity in Nude mice and immune compromised mice.
9. Increased activity of ribonucleotide reductase, RNA and DNA synthesis and decreased pyrimidine catabolism.

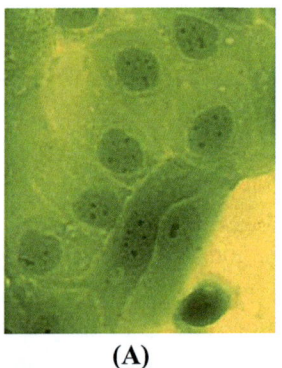
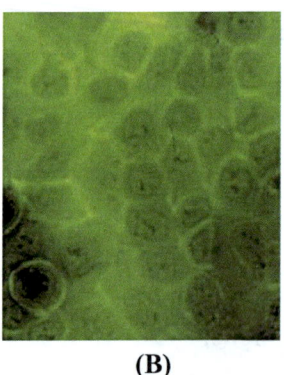

(A) (B)

Figure 7. Normal (a) vs. tumor (b) phenotype of the short columnar epithelium of the pancreatic ducts in *in vitro* culture that illustrates the several morphological changes associating dedifferentiation listed in the text.

Chapter 6

Tumor Recurrence and Metastasis

Tumor recurrence after surgical removal and/or chemo-radio-bio-therapies is due to new clonal expansion and/or residual disease evolution. Recurrent tumors are usually more aggressive than the first tumor reflecting the more complicated nature of newly acquired mutations. Rate of recurrence depends on type of the tumor and the extent of the field effect of the original tumor. Latency period, thus, differ from one tumor type to the other but also within same type of cancer.

The cancer death rate correlates its metastasizing capacity and invasive phenotype. Cancer metastasis is a complex and sequential network of cellular events involved in; the adhesion/anti-adhesion, invasion, migration, angiogenesis (angiogenic vs. antiangeogenic balance), establishment and proliferation of malignant cells from original site in distant secondary site. Secondary tumors are usually more aggressive than the primary ones. The pattern of metastasis (local vs. remote), metastasis route (lymph vs. blood) and the nature of the new location are specific for every type of cancer. This is due to the nature of interaction of the origin cells with surroundings and vasculatures; invasion mechanisms; nature of intravasation/extravasation adhesion mechanisms used, and the permissive nature at the new location.

The liver represents the most frequent visceral site of metastatic dissemination for virtually all types of neoplasms except brain primary cancer. Melanoma and lung cancers metastasize virtually to any organ. Whereas, large intestinal cancers generally spread to regional lymph nodes or to the liver via the portal venous circulation, distal rectum tumor spread through the paravertebral venous plexus and reach the lungs or supraclavicular lymph nodes without hepatic involvement. Metastasis of pancreatic cancer is typically to liver, peritoneum, and occasionally lung. Breast and prostate cancer have a propensity to metastasize to bone.

> Metastasis is promoted by oncogene-mediated signal transduction pathways that also suppress expression of invasion suppressor genes. Metastasis promoting genes include; AP-1; over-expression and/or mutational activation of c-erbB-1 (EGFR) and c-erbB-2; the chemokine receptor, CXCR4; the signaling tetraspanin CD151 and CO-029 receptors; mitogen-activated protein kinases; Rho family of small GTPases; G-protein-coupled receptors and genes encoding extracellular matrix processing proteases (urokinase plasminogen and specific matrix metalloproteases), adhesion proteins, and motility factors. Paradoxically, certain endogenous protease inhibitors such as PAI-1 and TIMP-1 appear to promote cancer metastasis rather than inhibiting the process. Some of these genes are also implicated in the early stages of tumorigenesis. Changes in the nature of glycation of the membrane glycoproteins and glycolipids of malignant cells are due to changes in activity of specific glycosyltransferases. Cells other than tumor cells contribute to the development of metastasis through their production of various pro-metastatic proteins such as collagen type I and Colony Stimulating Factor 1. This reflects the permissive environment for seeding of the metastasized cancer cells at specific location and not the others.

Metastasis suppressor genes inhibit cancer cell metastasis at any step of the metastatic cascade without affecting its tumorigenicity - despite the expression of oncogenes and metastasis-promoting activities in such cell. Their activity was established based on transfection and knock-out mouse studies. Metastasis suppressor genes exhibit a reduced expression in metastatic tumor cells due to mutations but most often due to epigenetic mechanisms. However, DNA methylation inhibitors have shown limited promise in derepressing them.

Metastasis suppressor genes work through regulating signaling pathways that modulate the aforementioned metastasis-promoting gene activities to control cell-cell and cell-matrix interactions. The first metastasis suppressor gene identified was nm23 (NDP kinase); others include: products of the signaling tetraspanin CD82/KAI1, mitogen-activated protein kinase-kinase 4, breast cancer metastasis suppressor-1, KiSS1, RHOGDI2, CRSP3, deleted in liver cancer 1 (DLC-1) gene (encodes a GTPase activating protein that acts as a negative regulator of the Rho family of small GTPases) and vitamin D3-upregulated protein/thioredoxin interacting protein.

Figure 8 illustrates an example for the multi-step nature of tumor evolution with progressive morphological dedifferentiation from normal colon cell to a precancerous state, to benign cancerous and finally to malignant state with ability to metastasize. This is due to and correlates the progressive acquisition of a series of somatic mutations/chromosomal abnormalities and epigenetic alterations leading to proto-oncogene activation (e.g., K-ras) and/or antioncogene inactivation (e.g., APC, DCC, and p53) - as biochemical dedifferentiation – as proposed by the Vogelstein model. Provided that the tumor microenvironment is permissive, in average, accumulation of 5-10 mutations is necessary for full transformation of a normal cell. Accumulation of such mutations/chromosomal abnormalities reflects the defective DNA mutation repair and/or genomic instability, i.e., hypermutable phenotype.

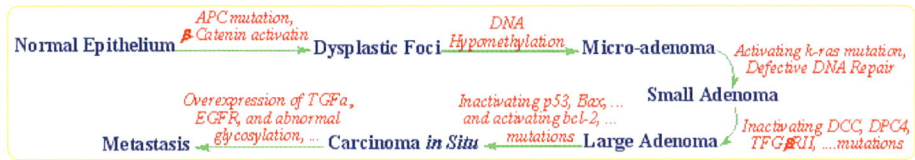

Figure 8. The sequential acquisition of activating and inactivation mutations, epigenetic changes and overexpression of different genes depicts the threshold phenomenon of the colorectal cancer pathogenesis as a model.

- Prostatic acid phosphatase (PAP) in prostate cancer, bone affection and myeloma.
- Alkaline phosphatase (ALP) in liver cancers, leukemia and sarcoma.
- Lactate dehydrogenase (LDH) in liver cancer, leukemia, lymphoma, and others.
- Neuron-Specific Enolase (NSE) in small cell lung cancer, neuroblastoma, melanoma, carcinoid disease, Wilms' tumor, and cancers of the thyroid, kidney, testicle, and pancreas.
- Proteases such as metalloproteases, plasmin and cathepsins in several cancers.
- Telomerase in several cancers.
- PSA is a serine protease specific for prostate cancer.
- Urokinase plasminogen activator (uPA) and plasminogen activator inhibitor 1 (PAI-1) for prognosis of newly diagnosed, node negative breast cancer.
- Thymidine kinase in several cancers.
- 5'-Nucleotidase, γ-glutamyl-transferase, aldolase and alcohol dehydrogenase in liver cancer.
- Sialyl-transferase in breast, lung and colon cancer.

Hormones and growth factors and their receptors and subreceptor effectors: such as,

- Calcitonin (CT) in medullary thyroid cancer.
- ACTH and bombesin in small cell lung cancer and parathormone (PTH) for non-small cell lung cancer.
- Prolactin and growth hormone in pituitary, renal and lung cancers.
- hCG trophoblastic tumors (invasive hydatidiform moles and choriocarcinomas) and germinal cell tumors (teratocarcinoma and embryonal cell carcinoma) of the ovaries and testes.
- Human epidermal growth factor receptor 2 (HER2) evaluation in breast cancer.
- Over production of growth factors such as TGFα, PDGF and IGF-I and -II in several tumors.
- K-ras mutation analysis for metastatic colorectal cancer and its response to therapy.
- Steroid hormone receptors (such as estrogen, progesterone and androgen receptors) in breast and prostate cancer.

Chapter 7

Tumor Markers

Tumor markers are alternative - to histopathology - indicators that predict future development, presence or future behavior of malignant tumors. They are secretory or tissue substances produced by tumor cells - at a level higher than normal and benign non-cancerous diseases - that reflect the presence of cancer genetic alterations or the host immunological reaction in response to tumor. Clinical chemistry and molecular biology techniques are used to identify these biomarkers in body fluids or tissue biopsies. These techniques mostly utilize monoclonal antibodies and include immunohistochemistry, flowcytometry, quantitative immunoassays (e.g., RIA and ELISA), DNA sequencing, cancer genetics and recombinant DNA processes.

Causes of the appearance of tumor markers:

a) Aberrant expression and/or modulation of normal antigens due to their mutation.
b) Expression of fetal antigens due to relative dedifferentiation of tumor cells with the production of specific tumor marker(s) by embryogenically related tissues.
c) Appearance of neoantigens (fusion proteins) due to abnormal chromosomal translocations.
d) The host immunological reaction in response to tumor, e.g., specific cytokines and antibodies against tumor markers.

Tumor biomarkers are classified on bases of the nature of their usage and application, and according to their biochemical and structural nature as follows. Classification of tumor markers according to their use:

a) *Markers for tumor differential diagnosis,* i.e., transformation-associating markers that discriminate cancer from other non-cancerous diseases.
b) *Markers for prediction of tumor prognosis*, i.e., markers that determine tumor grade, stage, progression and recurrence or response to therapy. However, half-life for clearance of the tumor markers from the blood after, e.g., resection or regressive treatment - varies and should be considered. It is, e.g., 12-24 hr for human chorionic gonadotropin (hCG), 2-3 days for Prostate-specific antigen (PSA), 5 days for alpha-fetoprotein (AFP).
c) *Markers for detection of metastasis or minimal residual disease.*
d) *Markers for screening and future susceptibility to cancer* in the general or a specific at-risk population, e.g., PSA + digital rectal examination for prostate cancer and genetic screening of BRCA1 and BRCA2 mutations for breast cancer.
e) *Tumor radioimmuno-localization,* using radiolabeled antibody directed against a tissue tumor marker.
f) *Tumor immunotherapy***,** using radiolabeled antibody or antibody conjugated to a cytotoxic agent or a cytotoxic agent activating enzyme directed against a tissue tumor marker.

Classification of tumor markers according to their nature:

1. **Tumor-related antigens:** They are either oncofetal or cancer-associated antigens;
    - *Oncofetal antigens:* They are expressed at high level in fetal serum and/or tissues and are reduced greatly or disappear after birth. They reappear with cancer due to reactivation of fetal genes. Examples include; carcinoembryonic antigen (CEA) in: liver, pancreas, lung, breast, ovary and gastrointestinal (GIT) cancers; carcinofetal ferritin in: liver cancer; and AFP in: choriocarcinoma in women, liver and gonadal non-seminomatous germ cell tumor.
    - *Cancer-associated antigens and/or mutations:* They are not expressed in normal cells and include:

- *Carbohydrate or mucin antigens (CA),* and GIT cancers, CA 125 in ovaria pancreas, liver, colon, breast, lung, and and some lymphomas; CA19-9 in col(pancreatic, stomach, hepatocellular c cancers; CA25 in Hairy cell leuken leukemia/lymphoma; CA 15-3 (CA 27-2 cancer but also ovary, lung, colon, stor uterus, and liver and prostate cancers; in stomach, kidney, lung, ovary, pancreas, ι CA50 in pancreas and colon cancers, antigen (MCA) in ovary and breast cance in pancreas, ovary, lung and GIT cancer sialic acid (LASA) in colorectal cancer.
- *Mutant forms of oncogenes and tumor* mutant p53 in several cancers including p21 (H, K, and N) in pancreas, colon, neuroblastoma, lymphoma and myel(retinoblastoma; Bcl2 in leukemia and neuroblastoma; c-myc in lung small cell and others.
- *Specific chromosomal translocations,* e, lymphoma and t(9,22) Philadelphia cl myelogenous leukemia.
- *Fusion protein or neoantigens,* e.g., tl protein, generated by the t(15;17) is obs acute promyelocytic leukemia and the B(generated by the t(9;22) is observed in myelogenous leukemia.
- *Squamous cell carcinoma antigen (SCC/* and head and neck cancers.
- *PSA* is a serine protease specific for prost;
- Monoclonal immunoglobulins and Benc(in multiple myeloma.

2. ***Enzymes:*** They include; fetal isozymes and ectopi produced enzymes, e.g.;

- Over production of angiogenic growth factors, e.g., FGFs and VEGF vs. reduced level of antiangiogenic factors, e.g., angiostatin and endostatin.
- Expression of c-erbB2 in breast, ovary and GIT cancers.
- Plasma catecholamines and their metanephrine and normetanephrine.in neuroblastoma and pheochromocytoma.

4. ***Markers of cellular proliferation:*** They are *in vivo* and/or *in vitro* biomarkers of cell proliferation, e.g.;
   - Higher than normal mitotic index.
   - Higher than normal thymidine labeling index or bromodeoxiuridine labeling index.
   - Higher than proliferating cell nuclear antigen (PCNA) expression.
   - Higher than Ki-67 (a subunit of DNA polymerase) nuclear expression.
   - DNA aneuploidy.
   - Higher than normal synthetic phase fraction.
   - Expression of nuclear matrix protein 22 (NMP22) in bladder cancer.
   - Higher than normal level of polyamines in CSF of brain cancer.

5. ***Markers of enhanced cellular turnover,*** such as,
   - Tissue polypeptide antigen (TPA) that is cytokeratins 8-, 18- and 19- derived oncofetal protein that has higher than normal level in breast, colon, ovary and bladder cancers.
   - Higher than normal level of the collagen degradation products, e.g., hydroxyproline.
   - Higher than normal level of the lipid-associating sialic acid in lung and GIT cancers.
   - Higher than normal level of the bladder tumor antigen (BTA), a basement membrane protein.

6. ***Miscellaneous:***
   - Genetic DNA alterations, e.g., loss of heterozygosity on the long arm of chromosome 18 that contain the Deleted in Colon Cancer (DCC) gene for colorectal cancer.
   - Acute phase reactants, e.g., acid glycoprotein, α_1-antitrypsin and β_2-microglobulin (β_2M) in B-cell lymphoma, lymphocytic leukemia and multiple myeloma.
   - C-peptide in insulinoma.
   - Ferritin in liver, lung and breast cancers and leukemia.

Specificity, Sensitivity and Predictivity of Tumor Markers

The accuracy of a screening test or its ability to discriminate a disease is reflected on four indices: sensitivity, specificity, positive predictive value, and negative predictive value. There are very few tumor markers - if any - that are 100% cancer specific, i.e., they do not lead to false positive diagnosis and 100% sensitive, i.e., they do not lead to false negative diagnosis.

Sensitivity of a tumor marker is its ability to detect a disease when it is present, i.e., the percentage of positive results among patients with the disease, typically it should be 100%. *Sensitivity = True +ve/(True +ve + False -ve) X 100.* For example, CEA is 72% sensitive (+ve only in 72% in late cancer colon patients) and is 20% sensitive (+ve only in 20% in early cancer colon patients).

Specificity of a tumor marker is its ability to tell that the disease is not present, i.e., the percentage of negative results among patients without the disease, typically it should be 100%. *Specificity = True -ve/(True -ve + False +ve) X 100.* For example, CEA is 80% specific among smokers since it is positive in 20% of normal smokers without cancer colon and is 97% specific among non-smokers since only 3% of them give positive result while having no disease.

Predictivity of the tumor marker is the percentage of true positive results among all positive cases or true negative among all negative cases. *Positive predictivity* is the proportion of persons who test positive among patients actually having the disease = *True +ve/(True +ve + False +ve) X 100*; Or, *= Prevalence X Sensitivity/(Prevalence X Sensitivity) + (1 - Prevalence)(1 - Specificity).* *Negative predictivity* is the proportion of persons who test negative among those without the disease = *True -ve/(True -ve + False -ve) X 100.* For example, CEA would be higher in patients attending colorectal clinic than in general population.

The sensitivity and specificity of a test are relatively independent of the underlying prevalence of the disease in the population screened, but the predictive values depend strongly on the prevalence of the disease. Predictivity is affect also by sensitivity of the test but more importantly by its specificity.

Nevertheless, tumor markers are invaluable mostly non-invasive and simple tools for cancer surveillance particularly when they are used as a battery of a number of markers for every type of cancer with their level exceeding a cut-off level specific for every marker. Examples for such batteries include: receptors of estrogen and progesterone, c-erbB2, CA 15-3, CA 27-29, MCA, MSA, and TPA for breast cancer; AFP, CEA, CA50, ferritin, ALP, and CA72-4 for liver cancer; and NSE, CYFRA21-1, CEA, ACTH, SCCA, CA15-3, TPA, ALP, and CA72-4 for lung cancer. Other diagnostic tools are usually utilized in combination including, histopathology, radiological imaging techniques and endoscopy.

Biomarkers For Exposure to Carcinogens

Animal bioassays and short-term laboratory tests for carcinogens are extremely useful to detect potential human carcinogens but their results are difficult to extrapolate quantitatively to humans. Therefore, alternative molecular cancer epidemiology biomarkers are used. Such molecular epidemiology biomarkers include:

- *Genetic and acquired host susceptibility;* such as increased specific cytochrome p450 activity; decreased specific glutathione-S-transferase activity; slow acetylators; efficiency of fidelity of DNA repair; susceptibility to tumor promoters; and, inherited mutations in antioncogenes.
- *Internal dose biomarkers (metabolism and tissue level of carcinogens and its metabolites in tissues and body fluids):* such as cotinine in serum and urine from cigarette smoke, 1-hydroxypyrene in urine from exposure to polycyclic aromatic hydrocarbons, aflatoxins in urine, and DDT in serum and body fat.
- *Biological effective dose biomarkers:* such as levels of covalent adduct formation with DNA and other macromolecules, e.g., hemoglobin and serum albumin for tobacco-specific nitrosamines.
- *Biomarkers of early cellular transformation and gene mutations* due to exposure to carcinogen such as DNA strand breaks, abnormal sister chromatid exchange, and chromosomal aberrations.
- *Biomarkers of tumor promoters* such as hormones, phorbol esters, aflatoxins, phenobarbital, tyrosine kinase activity and cell proliferation markers.

Chapter 8

Natural Inhibitors of the Multistage Carcinogenesis

Until the establishment of a real cure for cancer, the preventive approach would be more beneficial by using genetic selection and the natural dietary anticancer agents that include:

- *Chemicals that reduce the synthesis of carcinogens* in the body such as vitamin C that inhibits nitrosamine formation in the stomach.
- *Substances that reduce absorption of carcinogens* such as dietary fibers.
- *Chemicals that alter the metabolism of carcinogens* such as benzyl isothiocyanate from cruciferous vegetables such as cabbage, and selenium, vitamin E and carotenes as antioxidants from fruits and vegetables.
- *Chemicals that inhibit the covalent binding of carcinogens* to DNA such as elagic acid or flavonoids from fruits and vegetables.
- *Chemicals that inhibit tumor promotion and maintain cell differentiation* such as retinoids and other micronutrients from fruits and vegetables.
- *Inhibitors of farnesylation* of ras-p21 and hence its function by limonene from citrus fruits.
- *Natural anticancer chemicals without known mechanism* include organosulfur compounds (e.g., dimethyl allyl sulfinate) in garlic and onions, curcumin in turmeric, capsaicin in chili peppers, polyphenols in green tea and various protease inhibitors in several nutritional plant sources, e.g., tomatoes and potatoes.

Chapter 9

Biochemical Bases of Current Anticancer Treatments

Cancer is treated by many approaches as curative and/or palliative treatment (i.e., treatments that improves quality of patient life). They include surgical removal, and chemo- and radiotherapies.

- *Surgical "curative treatment"* by tumor excision combined with chemo- and/or radiotherapy using γ-emitting cobalt (^{60}Co) and cesium (^{137}Cs) or high energy electron beam, e.g., X-ray and linear accelerators. Surgical removal debulks most of the tumor and subsequent chemo- and/or radiotherapy will clean residual tumor cells. Dividing cancer/normal cells will be more susceptible to the DNA damaging effect of irradiation compared to non-dividing normal cells.
- *Chemotherapy* is cytotoxic treatments that interfere with cellular metabolism and/or cell division. Since cancer cells are ever-dividing with active metabolism for energy production and DNA synthesis they will be more susceptible to antimetabolic drugs, inhibitors of DNA synthesis, mutagens and inhibitors of cytoskeleton and spindle formation as compared to normal non-proliferating differentiated cells.

  Current chemotherapeutic drugs include:
  o Nucleotide analogs (mutagens), e.g., 5'-Fluorouracil and cytosine arabinoside.
  o Topoisomerase inhibitors, e.g., Doxorubicin.
  o DNA-alkylating agents (mutagens), e.g., melphalan and bleomycin.

- o Transmethylation inhibitors (antimetabolite), e.g., methotrexate.
- o Hormones and receptor antagonists, e.g., corticosteroids and antiestrogens.
- o DNA-strand breaking agents (mutagens), e.g., cisplatin.
- o Microtubule assembly inhibitors, e.g., vinblastine.
- o Inhibitors of nucleotides synthesis (antimetabolite), e.g., hydroxyurea.
- **Radiotherapy** induces single and double DNA strand breaks as direct effects. Also, indirect radiolysis of water into oxygen free radicals is more damaging to DNA as major cause of cell death. Cell death could be through necrosis or apoptosis depending on the dose and cell type. Cellular molecules other than DNA are subjected to damage by radiation, too.

The more promising rational recent chemotherapeutic approaches include:

- *Biotherapies,* e.g., antagonizing receptor antibodies such as herceptin - a humanized monoclonal antibody that binds and internalizes HER2/neu receptor to inhibit cell proliferation, and against EGFR; cytokines and antibodies labeled with cytotoxic drug/enzyme that activate a cytotoxic drug in target cells or radioactive element; potentiators of antitumor immunity, e.g., cancer vaccines against major proliferation-sustaining receptors and cytokines; and inhibitors of angiogenesis, e.g., angiostatin and endostatin.
- *Gene replacement therapy*, e.g., for p53, and, *gene disruption*, e.g., for ras-p21.
- *Apoptosis promoters*, e.g., gene disruption of Bcl2 and cell cycle inhibitors.
- *Differentiation stimulators*, e.g., retinoic acid and plant flavonoids.
- *Inhibitors of farnesyl transferase and telomerase activity.*
- *Glycosylation and protease inhibitors* that inhibit metastasis.
- *Photodynamic therapy.*
- *DNA demethylating drugs and histone deacetylase inhibitors.*
- *Prevention*, e.g., combating smoking, dietary fibers, and aspirin for colon cancer.
- *Protein kinase inhibitors.*

- *Genetic selection* before marriage and prophylactic surgical resection, e.g., breast resection for women at risk.

> Despite the erroneous basis of this approach in cancer treatment since it non-specifically targets dividing cells, the mercy of our Creator, made it productive in at least debulking tumor cells because normal cells that retain ability to repair mutations would escape mutagenic effects of such treatments and/or die by apoptosis. Thus, with current treatments (surgery and radio-, chemo-, and bio-therapies), >50% of patients diagnosed with cancer are cured!! (Better say treated)

Unfortunately, cancer cells develop mechanisms to withstand these types of treatments (Drug Resistance) through intrinsic and acquired pathways that include:

- *Poor drug absorption*, inadequate dosage, wrong schedule and drug interactions.
- *Tissue barriers* (e.g., brain and testes).
- *Kinetic resistance* (poor perfusion, large mass, high % of Go cells) and poor cellular uptake, poor drug activation and increased drug catabolism and efflux through resistance proteins.
- *Increase target enzyme*, altered target and decreased binding to target, and increased DNA repair.

Conclusion

The book highlighted the outmost importance of cancer prevention as a priority in an era of environmental pollution with so many carcinogens and cancer promoters - with proper understanding of their mechanism of action. The hereditary predisposition should be addressed as serious as the other causes - taking the successful approach to eradicate specific inborn errors of metabolism and hemoglobinopathies as a model. Understanding the initial pathogenetic alteration into cancer transformation as a dedifferentiation process that includes the uncontrolled cell division due to tumor suppressor genes/oncogenes imbalance is crucial. Moreover, it is important to consider cancer transformation as a threshold phenomenon - due to cumulative inactivation of tumor suppressor gene and/or activation of oncogenes - that affects the body cells but erupts only in a cell of a metabolically favorable tissue - as the weakest spot of the decaying pipe. This explains the cancer subsequent aggression and metastasis. This book is hoped to establish a proper understanding for cancer upon which a proper treatment approaches could be envisioned in the near future.

References

Cancer Etiology and Carcinogenesis:

Buchschacher Jr. GL and Wong-Staal F. Etiology of cancer: RNA viruses. In, Cancer: Principles and Practice of Oncology, 7th Edition (DeVita, Jr., VT, Hellman S and Rosenberg SA; editors); 2005; Part 2, Chapter, 7, Section 1, pp165-173. Lippincott Williams and Wilkins, Philadelphia, USA.

Calvo KR, Petricoin III, EF and Liotta LA. Genomics and proteomics. In, Cancer: Principles and Practice of Oncology, 7th Edition (DeVita, Jr., VT, Hellman S and Rosenberg SA; editors); 2005; Part 1, Chapter, 2, pp51-72. Lippincott Williams and Wilkins, Philadelphia, USA.

Chan TA, Glockner S, Yi JM, et al. Convergence of mutation and epigenetic alterations identifies common genes in cancer that predict for poor prognosis. PLoS. Med. 2008; 5(5):e114.

Colditz GA and Fisher LB. Etiology of cancer: Tobacco use. In, Cancer: Principles and Practice of Oncology, 7th Edition (DeVita, Jr., VT, Hellman S and Rosenberg SA; editors); 2005; Part 2, Chapter, 9, pp193-199. Lippincott Williams and Wilkins, Philadelphia, USA.

El-Metwally TH, Ali HA and Anwar MM. Role of mutant p53, p21-ras and EGF receptor compared to conventional tumor markers in colorectal cancer in upper Egyptian patients. *The Egyptian J. Biochemistry and Molecular Biol.*; 2002, 20 (special issue):263-164.

El-Metwally TH. Transformation – normalizing-redifferentiation – apoptosis sequence and the role of the mitochondria in the retinoid-pancreatic cancer model: Is it obligatory? In: Cell Differentiation Research Developments

(Ivanova LB, editor), 2007, Chapter XI, pp 235-47 Nova Science Publishers, Inc., Hauppauge, NY 11788.

Fenton RG and Longo DL. Cancer cell biology and angiogenesis. In, Harrison's Principles of Internal Medicine (Kasper DL et al., editors), 16th edition, 2005; Part V, Chapter 69,, pp453-64. McGraw-Hill Medical Publishing Division, NY.

Howley PM, Ganem D and Kieff E. Etiology of cancer: DNA viruses. In, Cancer: Principles and Practice of Oncology, 7th Edition (DeVita, Jr., VT, Hellman S and Rosenberg SA; editors); Part 2, Chapter, 7, Section 2, pp1673-184. Lippincott Williams and Wilkins, 2005; Philadelphia, USA.

Morin PJ, Trent JM, Collins FS, Vogelstein B. Cancer genetics. In, Harrison's Principles of Internal Medicine (Kasper DL et al., editors), 16th edition, 2005; Part V, Chapter 68, pp447-53. McGraw-Hill Medical Publishing Division, NY.

Neal AJ, Hoskin PJ (editors). Pathogenesis of cancer. In, Clinical oncology: Basic principles and practice, 3rd edition, 2003; Chapter, 1, pp1-7. Arnold, London, UK.

Porter PL. Molecular markers of tumor initiation and progression. *Curr. Opin. Genet. Dev.*; 2001; 11(1):60-3.

Sack J and Rothman JM. Colorectal Cancer: Natural History and Management. *Hospital Physician*, 2000; 64-73.

Schmied BM, Ulrich AB, Matsuzaki H, El-Metwally TH, et al. Biological instability of pancreatic cancer xenografts in the nude mouse. *Carcinogenesis*, 2000; 21(6):1121-7.

Souhami R and Tobias J (editors). Biology of cancer. In, Cancer and its management; 5th edition, 2005; chapter 3, pp23-41. Blackwell publishing, Oxford, UK.

Souhami R and Tobias J (editors). Epidemiology, cure, treatment trials and screening. In, Cancer and its management; 5th edition, 2005; chapter 2, pp7-22. Blackwell publishing, Oxford, UK.

Ullrich RL. Etiology of cancer: Physical factors. In, Cancer: Principles and Practice of Oncology, 7th Edition (DeVita, Jr., VT, Hellman S and Rosenberg SA; editors); 2005; Part 2, Chapter, 10, pp201-215. Lippincott Williams and Wilkins, Philadelphia, USA.

Yuspa SH and Shields PG. Etiology of cancer: Chemical factors. In, Cancer: Principles and Practice of Oncology, 7th Edition (DeVita, Jr., VT, Hellman S and Rosenberg SA; editors); 2005; Part 2, Chapter, 8, pp185-191. Lippincott Williams and Wilkins, Philadelphia, USA.

Chemical Carcinogens

Bard D, Barouki R, Benhamou S, Benichou J, Clavel J, Jougla E, and Launoy G. Cancer A methodological approach for studying the link between cancer and the environment: Collective Expert Report; 2006, pp1-86. Inserm (National Institute for health and medical research), Paris, France.

Bassil KL, Vakil C, Sanborn M, Cole DC, Kaur JS, and Kerr KJ. Cancer health effects of pesticides: Systematic review. *Can. Fam. Physician.*; 2007; 53(10):1704-11.

Bard D, Barouki R, Benhamou S, Benichou J, Clavel J, Jougla E, and Launoy G. Cancer A methodological approach for studying the link between cancer and the environment: Collective Expert Report; 2006, pp1-86. Inserm (National Institute for health and medical research), Paris, France.

Grosse Y, Baan R, Straif K, et al. A review of human carcinogens-Part A: pharmaceuticals. *Lancet Oncol.*; 2009; 10(1):13-4.

Palut D, Kostka G, Adamczyk M. Molecular mechanisms of chemically induced carcinogenesis. Rocz Panstw Zakl Hig.; 1998; 49(1):35-54.

Tumor Suppressor Genes

Banerjee HN and Verma M. Tumor Suppressor Genes; Chapter I, pp. 19-33. In, Tumor Suppressor Genes (Polinsky KR, Editor), 2007. Nova Science Publishers, Inc., Hauppauge, NY 11788.

Espinozia LA and Crotti LB. Roles of PARP-1 and p53 in the Maintenance of Genome Integrity; Chapter II, pp. 35-50. In, Tumor Suppressor Genes (Polinsky KR, Editor), 2007. Nova Science Publishers, Inc., Hauppauge, NY 11788.

Guan Y-S, He Q, and Hu Y. The p53 Gene; Chapter V, pp. 95-125. In, Tumor Suppressor Genes (Polinsky KR, Editor), 2007. Nova Science Publishers, Inc., Hauppauge, NY 11788.

Hansen JE, Nishimura RN, and Heinze E. Fv-p53 Protein Therapy Presents an Alternative to Gene Therapy; Chapter III, pp. 51-63. In, Tumor Suppressor

Genes (Polinsky KR, Editor), 2007. Nova Science Publishers, Inc., Hauppauge, NY 11788.
Javier RT, Butel JS. The history of tumor virology. Cancer Res.; 2008; 68(19):7693-706.
Kumagai T. Reactivation of Epigenetically Silenced Tumor Suppressor Genes as a Molecular Targeted Therapy; Chapter IV, pp. 65-94. In, Tumor Suppressor Genes (Polinsky KR, Editor), 2007. Nova Science Publishers, Inc., Hauppauge, NY 11788.
Lacroix M. An Update on Tumor Suppressor Genes in Breast Cancer; Chapter VII, pp. 177-251. In, Tumor Suppressor Genes (Polinsky KR, Editor), 2007. Nova Science Publishers, Inc., Hauppauge, NY 11788.
Madhavan J, Ganesh A, Kumaramanickavel G. Retinoblastoma: from disease to discovery. *Ophthalmic Res.*; 2008; 40(5):221-6.
Mastrangelo D, De Francesco S, Di Leonardo A, Lentini L, and Hadjistilianou T. The retinoblastoma paradigm revisited. *Med. Sci. Monit.*; 2008; 14(12):RA231-40.
Qayum I and Ashraf M. DNA Methylation and Tumor Suppression; Chapter VIII, pp. 253-279. In, Tumor Suppressor Genes (Polinsky KR, Editor), 2007. Nova Science Publishers, Inc., Hauppauge, NY 11788.
Sábado Alvarez C. Molecular biology of retinoblastoma. *Clin. Transl. Oncol.*; 2008; 10(7):389-94.
Wang W-D, Li R, and Chen Z-T. Tumor Suppressor Gene P53 Molecular Biology Prognostic and Therapeutic Strategies for Human Cancers; Chapter VI, pp. 127-176. In, Tumor Suppressor Genes (Polinsky KR, Editor), 2007. Nova Science Publishers, Inc., Hauppauge, NY 11788.
Whibley C, Pharoah PD, Hollstein M. p53 polymorphisms: cancer implications. *Nat. Rev. Cancer*; 2009; 9(2):108-22.

Oncogenes

Frenzel A, Grespi F, Chmelewskij W, Villunger A. Bcl2 family proteins in carcinogenesis and the treatment of cancer. Apoptosis. 2009 - ahead of printing.
Kuttler F, Mai S. c-Myc, Genomic Instability and Disease. *Genome Dyn.*; 2006; 1:171-90.

McCormick F. Ras-related proteins in signal transduction and growth control. *Mol. Reprod. Dev.*; 1995; 42(4):500-6.

Pincus MR, Brandt-Rauf PW, Michl J, Carty RP, Friedman FK. ras-p21-induced cell transformation: unique signal transduction pathways and implications for the design of new chemotherapeutic agents. *Cancer Invest.*; 2000; 18(1):39-50.

Prochownik EV. c-Myc: linking transformation and genomic instability. *Curr. Mol. Med.*; 2008; 8(6):446-58.

Ramakrishna G, Sithanandam G, Cheng RY, Fornwald LW, Smith GT, Diwan BA, Anderson LM. K-ras p21 expression and activity in lung and lung tumors. *Exp Lung Res.*; 2000; 26(8):659-71.

Shinohara N, Koyanagi T. Ras signal transduction in carcinogenesis and progression of bladder cancer: molecular target for treatment? *Urol. Res.*; 2002; 30(5):273-81.

Thomadaki H and Scorilas A. BCL2 family of apoptosis-related genes: functions and clinical implications in cancer. *Crit. Rev. Clin. Lab. Sci.*; 2006;43(1):1-67.

Yoshimatsu K. Inhibitors of isoprenylation of ras p21. Gan To Kagaku Ryoho. ;1997; 24(11):1495-502.

Oxidative Stress and Cancer

Gogvadze V, Orrenius S, Zhivotovsky B. Mitochondria as targets for chemotherapy. Apoptosis. 2009 - ahead of printing.

Khandrika L, Kumar B, Koul S, Maroni P, Koul HK. Oxidative stress in prostate cancer. Cancer Lett. 2009 - ahead of printing.

Invasion, Metastasis and Microenvironment

Ahmed F, Steele JC, Herbert JM, Steven NM, Bicknell R. Tumor stroma as a target in cancer. *Curr Cancer Drug Targets*; 2008; 8(6):447-53.

Chan DA, Giaccia AJ. Hypoxia, gene expression, and metastasis. *Cancer Metastasis Rev.* 2007; 26(2):333-9.

Chiang AC, Massagué J. Molecular basis of metastasis. *N. Engl. J. Med.*; 2008; 359(26): 2814-23.

Fidler IJ, Langley RR, Kerbel RS and Ellis LM. Angiogenesis. In, Cancer: Principles and Practice of Oncology, 7th Edition (DeVita, Jr., VT, Hellman S and Rosenberg SA; editors); 2005; Part 1, Chapter, 5, pp129-137. Lippincott Williams and Wilkins, Philadelphia, USA.

Fidler IJ. Critical determinants of metastasis. *Semin Cancer Biol.*; 2002; 12(2):89-96.

Francis P, Namløs HM, Müller C, et al. Diagnostic and prognostic gene expression signatures in 177 soft tissue sarcomas: hypoxia-induced transcription profile signifies metastatic potential. *BMC Genomics*; 2007; 8:73.

Glinsky GV, Berezovska O, and Glinskii AB. Microarray analysis identifies a death-from-cancer signature predicting therapy failure in patients with multiple types of cancer. *J. Clin. Invest.*; 2005; 115(6): 1503–21.

Pantel K, Brakenhoff RH. Dissecting the metastatic cascade. *Nat. Rev. Cancer*; 2004; 4(6):448-56.

Rennstam K and Hedenfalk I. High-throughput genomic technology in research and clinical management of breast cancer. Molecular signatures of progression from benign epithelium to metastatic breast cancer. *Breast Cancer Res.*; 2006; 8(4): 213.

Schedin P and Elias A. Multistep tumorigenesis and the microenvironment. *Breast Cancer Res.*; 2004; 6:93-101.

St Hill CA, Farooqui M, Mitcheltree G, Gulbahce HE, Jessurun J, Cao Q, Walcheck B. The high affinity selectin glycan ligand C2-O-sLex and mRNA transcripts of the core 2 beta-1,6-N-acetylglusaminyltransferase (C2GnT1) gene are highly expressed in human colorectal adenocarcinomas. *BMC Cancer.* 2009;9(1):79.

Stetler-Stevenson WG. Invasion and Metastases. In, Cancer: Principles and Practice of Oncology, 7th Edition (DeVita, Jr., VT, Hellman S and Rosenberg SA; editors); Part 1, Chapter, 4, pp113-127. Lippincott Williams and Wilkins, 2005; Philadelphia, USA.

Steeg PS. Tumor metastasis: mechanistic insights and clinical challenges. *Nat. Med.*; 2006; 12(8):895-904.

Tutt A, Wang A, Rowland C, et al. Risk estimation of distant metastasis in node-negative, estrogen receptor-positive breast cancer patients using an RT-PCR based prognostic expression signature. *BMC Cancer* ; 2008; 8:339.

Yoshida BA, Sokoloff MM, Welch DR, Rinker-Schaeffer CW. Metastasis-Suppressor Genes: a Review and Perspective on an Emerging Field. *J. National Cancer Institute*; 2000; 92(21):1717-30.

Zhang B, Cao X, Liu Y, et al. Tumor-derived matrix metalloproteinase-13 (MMP-13) correlates with poor prognoses of invasive breast cancer. *BMC Cancer*; 2008; 8:83.

Tumor Markers

Brooks M. Breast cancer screening and biomarkers. *Methods Mol. Biol.*; 2009; 472:307-21.

Kensler TW, Davidson NE, Groopman JD, Muñoz A. Biomarkers and surrogacy: relevance to chemoprevention. *IARC Sci. Publ.*; 2001; 154:27-47.

Owen RW. Biomarkers in colorectal cancer. IARC Sci. Publ.; 2001;.154:101-11.

Saran A, Kumar U, Shahi SK, Rai HS, Jaiswal CP. Evaluation of tumor markers in carcinoma breast. *Indian J. Pathol. Microbiol.*; 2000;43(4):437-40.

Souhami R and Tobias J (editors). Staging of tumors. In, Cancer and its management; 5th edition, 2005;chapter 4, pp42-56. Blackwell publishing, Oxford, UK.

Volpe A, Racioppi M, D'Agostino D, et al. Bladder tumor markers: a review of the literature. *Int. J. Biol. Markers*; 2008; 23(4):249-61.

Cancer Prevention

Brawley OW, Kramer BS. Prevention and early detection of cancer. In, Harrison's Principles of Internal Medicine (Kasper DL et al., editors), 16th edition, 2005; Part V, Chapter 67, pp 441-7. McGraw-Hill Medical Publishing Division, NY.

Greenwald P. Cancer prevention: Diet and chemopreventive agents: Dietary carcinogens. In, Cancer: Principles and Practice of Oncology, 7th Edition (DeVita, Jr., VT, Hellman S and Rosenberg SA; editors); 2005; Part 3, Chapter, 20, Section 5, pp536-540. Lippincott Williams and Wilkins, Philadelphia, USA.

Koh HK and Geller AC. Cancer prevention: Preventing tobacco-related cancer. In, Cancer: Principles and Practice of Oncology, 7th Edition (DeVita, Jr., VT,

Hellman S and Rosenberg SA; editors); 2005; Part 3, Chapter, 19, pp493-505. Lippincott Williams and Wilkins, Philadelphia, USA.

Mayne ST and Lippman SM. Cancer prevention: Diet and chemopreventive agents: Retinoids, carotenoids and micronutrients. In, Cancer: Principles and Practice of Oncology, 7th Edition (DeVita, Jr., VT, Hellman S and Rosenberg SA; editors); 2005; Part 3, Chapter, 20, Section 4, pp521-533. Lippincott Williams and Wilkins, Philadelphia, USA.

Michels KB. Cancer prevention: Diet and chemopreventive agents: Dietary fibers. In, Cancer: Principles and Practice of Oncology, 7th Edition (DeVita, Jr., VT, Hellman S and Rosenberg SA; editors); 2005; Part 3, Chapter, 20, Section 2, pp514-517. Lippincott Williams and Wilkins, Philadelphia, USA.

Michels KB. Cancer prevention: Diet and chemopreventive agents: Fruit and vegetable consumption. In, Cancer: Principles and Practice of Oncology, 7th Edition (DeVita, Jr., VT, Hellman S and Rosenberg SA; editors); 2005; Part 3, Chapter, 20, Section 3, pp518-520. Lippincott Williams and Wilkins, Philadelphia, USA.

The Twenty-First Aspen Cancer Conference: Mechanisms of Toxicity, Carcinogenesis, Cancer Prevention and Cancer Therapy, July 16–19, 2006; and Cancer Prevention: Life Style or Nutrition? July 20–21, 2006, The Gant Conference Center, Aspen, Colorado. Toxicologic Pathology; 2006; 34:968-1018.

Thun MJ and Henley SJ. Cancer prevention: Cyclooxygenase inhibitors. In, Cancer: Principles and Practice of Oncology, 7th Edition (DeVita, Jr., VT, Hellman S and Rosenberg SA; editors); 2005; Part 3, Chapter, 20, Section 6, pp541-8. Lippincott Williams and Wilkins, Philadelphia, USA.

Willett WC. Cancer prevention: Diet and chemopreventive agents: Dietary fat. In, Cancer: Principles and Practice of Oncology, 7th Edition (DeVita, Jr., VT, Hellman S and Rosenberg SA; editors); 2005; Part 3, Chapter, 20, Section 1, pp507-511. Lippincott Williams and Wilkins, Philadelphia, USA.

Cancer Treatment

Choudhury A, Singh RK, Moniaux N, El-Metwally TH, Aubert JP, Batra SK. Retinoic Acid-dependent transforming growth factor-beta2-mediated induction of MUC4 mucin expression in human pancreatic tumor cells follows retinoic acid receptor-alpha signaling pathway. *J. Biol. Chem.*; 2000; 275(43): 33929-36.

Ding XZ, Kuszynski CA, El-Metwally TH, Adrian TE. Lipoxygenase inhibition induced apoptosis, morphological changes, and carbonic anhydrase expression in human pancreatic cancer cells. *Biochem. Biophys. Res. Commun.* (BBRC); 1999; 266, 392-9.

EL-Metwally TH and Adrian TE. Optimization of treatment conditions for studying the anticancer effects of retinoids using pancreatic cancer as a model. *BBRC*; 1999; 257, 596-603.

El-Metwally TH and Hameed DA. The effectiveness of retinoic acid treatment in bladder cancer: Impact on recurrence, survival and TGFα and VEGF as end-point biomarkers. *Cancer Biology and Theapy*; 2008; 7(1); 1-9.

El-Metwally TH and Pour PM. The Retinoid Induced Pancreatic Cancer Redifferentiation-Apoptosis Sequence and the Mitochondria: A Suggested Obligatory Sequence of Events, Review. *J Of Pancreas*; 2007; 8(3):1001-1011.

El-Metwally TH, Hussein MR, Abd-El-Ghaffar SKh, Abo-El-Naga MM, Ulrich AB, Pour PM. Retinoic acid can induce markers of endocrine transdifferentiation in pancreatic ductal adenocarcinoma: preliminary observations from an in vitro cell line model. *J. Clin. Pathol.*; 2006;59(6):603-10.

El-Metwally TH, Hussein MR, Pour PM, Kuszynski CA, Adrian TE. High concentrations of retinoids induce differentiation and late apoptosis in pancreatic cancer cells in vitro. *Cancer Biol. Ther.*; 2005; 4(5):602-11.

El-Metwally TH, Hussein MR, Pour PM, Kuszynski CA, Adrian TE. Natural retinoids inhibit proliferation and induce apoptosis in pancreatic cancer cells previously reported to be retinoid resistant. *Cancer Biol. Ther.*; 2005;4(4):474-83.

Hahn SM and Glatstein E. Principle of radiation therapy. In, Harrison's Principles of Internal Medicine (Kasper DL et al., editors), 16th edition; 2005; Part V, Chapter 71, pp482-9. McGraw-Hill Medical Publishing Division, NY.

Rogers LJ, Eva LJ, Luesley DM. Vaccines against cervical cancer. *Methods Mol. Biol.*; 2009;472:25-56.

Sausville EA and Longo DL. Principles of cancer treatment: Surgery, chemotherapy and biologic therapy. In, Harrison's Principles of Internal Medicine (Kasper DL et al., editors), 16th edition; 2005; Part V, Chapter 70, pp464-82. McGraw-Hill Medical Publishing Division, NY.

Souhami R and Tobias J (editors). Radiotherapy. In, Cancer and its management; 5th edition, chapter 5, pp57-75. Blackwell publishing, Oxford, UK.

Souhami R and Tobias J (editors). Systemic treatment of cancer. In, Cancer and its management; 5th edition, 2005; chapter 6, pp76-107. Blackwell publishing, Oxford, UK.

Zhang J, Yang PL, Gray NS. Targeting cancer with small molecule kinase inhibitors. *Nat. Rev. Cancer*; 2009;9(1):28-39.

Index

A

accuracy, 52
acid, x, 16, 23, 25, 28, 50, 51, 55, 58, 71
ACTH, 50, 52
activation, 5, 7, 10, 11, 18, 23, 25, 26, 27, 28, 29, 30, 36, 38, 41, 45, 59, 61
acute myeloid leukemia, 7, 33
acute promyelocytic leukemia, 49
addiction, 37
adenocarcinoma, 31, 71
adenovirus, 34
adhesion, 31, 33, 43
ADP, 20
agar, 41
age, 7, 9, 36, 38
agent, 20, 48
aggression, 61
aggressive behavior, xi, 26
albumin, 42
alcohol, 15, 37, 50
aldolase, 50
alkylation, 28
allele, 20, 27, 35
alpha-fetoprotein, 48
alternative, 28, 34, 47, 53
amines, 11, 16, 37
amino acids, 35

androgen, 50
androgens, 13, 15
anemia, 7, 33
aneuploidy, 30, 42, 51
angiogenesis, 9, 33, 43, 58, 64
angioma, 33
angiosarcoma, 16
anhydrase, 42, 71
animals, 1
antibody, 48
anticancer drug, 26
antigen, 25, 34, 48, 49, 51
antioxidant, 19
antitumor, 6, 58
APC, 29, 33, 37, 38, 45
apoptosis, x, xi, 3, 5, 6, 8, 9, 17, 20, 26, 30, 31, 33, 34, 35, 39, 42, 58, 63, 67, 71
apoptosis pathways, 30
apoptotic mechanisms, 24
arabinoside, 57
ARC, 69
arrest, 6, 34, 35
asbestos, 10, 12
asymptomatic, 38
ataxia, 32
atoms, 12

Index

ATP, 8, 19
atrophy, 6
autoimmune disease, x
autosomal dominant, 6, 36
autosomal recessive, 6, 7, 36

B

bacteria, 9
barriers, 59
basal cell carcinoma, 30
basement membrane, 51
batteries, 52
behavior, x, 1, 47
benign, 13, 31, 45, 47, 68
benign tumors, 13
benzo(a)pyrene, 11, 12, 13
beverages, 15
bile duct, 49
binding, ix, 26, 27, 28, 34, 35, 55, 59
biomarkers, 47, 48, 51, 53, 69, 71
birth, x, 48
bladder, 15, 16, 24, 37, 51, 67, 71
blocks, 6, 33
blood, 19, 43, 48
blood supply, 19
body fat, 53
body fluid, 47, 53
bonds, 18
bone marrow, x, 1
brain, 26, 27, 31, 43, 51, 59
breast cancer, 7, 27, 29, 31, 32, 33, 37, 38, 44, 48, 49, 50, 51, 52, 68, 69
breast carcinoma, 23
Burkitt's lymphoma, 29

C

cabbage, 55
cancer, ix, x, 1, 4, 5, 6, 8, 9, 10, 12, 13, 15, 16, 17, 18, 19, 21, 23, 24, 26, 27, 29, 30, 31, 32, 33, 35, 36, 37, 38, 41, 42, 43, 44, 47, 48, 49, 50, 51, 52, 53, 55, 57, 58, 59, 61, 63, 64, 65, 66, 67, 68, 69, 71, 72
cancer cells, x, 1, 19, 24, 27, 30, 41, 42, 57, 59
cancer screening, 69
carcinoembryonic antigen, 48
carcinogen, 7, 11, 12, 17, 19, 36, 53
carcinogenesis, 3, 12, 17, 18, 19, 65, 66, 67
carcinoma, 6, 22, 23, 24, 29, 32, 49, 50, 69
carotenoids, 70
cartilage, 2
casein, 34
catabolism, x, 42, 59
catecholamines, 51
causation, 10
cell, ix, x, xi, 1, 3, 5, 6, 8, 11, 12, 13, 17, 19, 20, 21, 22, 23, 24, 25, 26, 27, 28, 29, 30, 31, 32, 33, 34, 35, 37, 41, 44, 45, 48, 49, 50, 51, 53, 55, 57, 58, 61, 64, 67, 71
cell cycle, 3, 6, 27, 29, 30, 31, 32, 34, 35, 37, 58
cell death, 20, 21, 29, 58
cell line, 71
cell metabolism, 30
cell surface, 41
cellular signaling pathway, 14
cervical cancer, 22, 25, 71
cervix, 49
cesium, 57
charring, 15
chemoprevention, 38, 69
chemopreventive agents, 69, 70
chemotherapeutic agent, 67
chemotherapy, 11, 34, 67, 71
chicken, 28
children, 38
choriocarcinoma, 48
chromosomal abnormalities, 17, 45
chromosomal instability, 7, 24, 33
chromosome, 7, 25, 35, 49, 51
chronic lymphocytic leukemia, 34
chronic myelogenous, 25, 49
cigarette smoke, 53

circulation, 43
cobalt, 57
coding, 7, 21
codon, 11
collagen, 51
colon, x, xi, 1, 6, 7, 23, 24, 27, 29, 32, 35, 37, 38, 45, 49, 50, 51, 52, 58
colon cancer, 7, 27, 29, 32, 37, 49, 50, 58
colonoscopy, 38
colorectal adenocarcinoma, 68
colorectal cancer, 25, 31, 32, 33, 34, 45, 49, 50, 51, 63, 69
communication, 41
components, 7, 33
compounds, 13, 17, 19, 55
concordance, 38
configuration, 25
conjugation, 28
connective tissue, 2, 35
consumption, 70
contraceptives, 15
control, ix, 22, 24, 25, 34, 37, 44, 67
conversion, xi, 8
cornea, 6
corticosteroids, 58
cotinine, 53
CSF, 23, 51
culture, 21, 42
cyclins, 30, 36
cyclophosphamide, 11
cytochrome, 53
cytochrome p450, 53
cytokines, 47, 58
cytoplasm, 41
cytosine, 57
cytoskeleton, 33, 57

D

death, x, 1, 15, 29, 34, 43, 58, 68
death rate, 43
deaths, 9, 15, 16

defects, x
deficiency, 6
degradation, 6, 20, 32, 33, 34, 51
density, 33, 41
deoxyribose, 18
detection, 48, 69
developed countries, 9
diet, 38
dietary fiber, x, 55, 58
differential diagnosis, 48
differentiation, ix, x, xi, 1, 3, 9, 18, 20, 21, 25, 28, 30, 33, 42, 55, 71
dimerization, 28
dioxin, 9, 13, 19
disease progression, 34
disorder, 23
division, ix, x, 1, 22, 30, 35, 57, 61
DNA, x, xi, 5, 6, 7, 9, 10, 11, 12, 13, 17, 18, 19, 21, 23, 24, 25, 27, 28, 30, 31, 32, 34, 35, 36, 37, 42, 44, 45, 47, 51, 53, 55, 57, 58, 59, 64, 66
DNA damage, xi, 6, 8, 10, 34, 35
DNA polymerase, 51
DNA repair, 6, 7, 9, 17, 24, 27, 32, 35, 37, 53, 59
DNA sequencing, 47
DNA strand breaks, 10, 53, 58
dopamine, 37
dosage, 27, 59
double bonds, 18
drug interaction, 59
drug resistance, 3, 26
drugs, 57, 58

E

electromagnetic waves, 9, 10
electron, 12, 57
ELISA, 47
elongation, 33
embryogenesis, x
endocrine, 71

endoscopy, 53
energy, 8, 9, 10, 18, 20, 57
environment, xi, 10, 65
environmental factors, 38
enzymes, 37, 49
epidemiology, 53
epithelium, x, xi, 42, 68
Epstein-Barr virus, 21, 24, 25
equilibrium, 3
esophageal cancer, 38
esophagus, 15
estrogen, 50, 52, 68
etiology, 22
euchromatin, 41
evolution, 4, 43, 45
excision, 6, 7, 32, 37, 57
exons, 21, 26, 28, 35
exonuclease, 7
exposure, 9, 13, 36, 37, 53
extravasation, 43

F

failure, 24, 68
false negative, 52
false positive, 52
family, 7, 8, 23, 29, 34, 36, 38, 44, 66, 67
family history, 38
family members, 7, 34
fat, 2, 15, 70
fermentation, x
ferritin, 48, 52
fetus, 36
fibers, 38, 70
fibrosarcoma, 22
fidelity, 53
fish, 15
flavonoids, x, 38, 55, 58
food, 15
food additives, 15
frameshift mutation, 33
France, 65, 66

free radicals, 8, 10, 13, 18, 19, 58
fruits, 55
functional changes, 22, 26
fusion, 23, 25, 47, 49

G

GDP, 24, 27
gene, 5, 6, 7, 11, 21, 22, 23, 24, 25, 26, 27, 28, 29, 30, 31, 32, 33, 35, 37, 38, 44, 51, 53, 58, 61, 67, 68
gene amplification, 5, 23, 30
gene expression, 22, 25, 67, 68
generation, 8, 30
genes, ix, x, 3, 4, 6, 7, 9, 18, 20, 21, 22, 25, 27, 28, 29, 32, 34, 36, 37, 38, 44, 45, 48, 49, 61, 63, 67
genetic alteration, 3, 35, 36, 47
genetic factors, 15, 38
genetic screening, 48
genetic testing, 38
genetics, 47, 64
genome, 5, 21, 24, 26, 34
genomic instability, 6, 30, 35, 45, 67
genotype, 37
germline mutations, 37
gland, 13
glioma, 22, 26
glucose, 8, 31
glutathione, 37, 53
glycolysis, 8, 30, 41
glycoproteins, 41
government, vi
grades, x
groups, 7, 32
growth, 1, 3, 19, 21, 22, 26, 27, 29, 33, 34, 41, 42, 50, 51, 67
growth factor, 21, 22, 26, 27, 29, 41, 42, 50, 51
growth hormone, 50
growth rate, 3
guanine, 7, 11, 12

guardian, 34
guidance, 31

H

half-life, 28, 35, 48
hazards, x
HBV, 16
head and neck cancer, 49
health, 1, 65
health effects, 65
heart disease, 1
hemoglobin, 53
hemoglobinopathies, 61
hepatitis, 15, 21
hepatocellular cancer, 49
hepatocellular carcinoma, 15, 22, 25, 29
hepatocytes, x, 42
heterochromatin, 41
histone, 25, 28, 58
HIV, 6, 16
HLA, 37
hormone, 24, 50
host, 24, 47, 53
human chorionic gonadotropin, 48
human immunodeficiency virus, 5, 21
human papilloma virus, 22, 34
hybrid, 25
hydatidiform mole, 50
hydrocarbons, 16
hydrogen, 20
hydrogen peroxide, 20
hydrolysis, 19
hydroxyl, 18
hypothesis, 35
hypoxia, 19, 34, 68

I

immunity, 9, 37, 58
immunodeficiency, 6, 32
immunoglobulin, 24, 25, 29, 31

immunoglobulins, 49
immunohistochemistry, 47
immunotherapy, 48
imprinting, 27
in vitro, 26, 42, 51, 71
in vivo, 19, 51
incidence, 1, 22
independence, 42
indicators, 47
indices, 52
induction, 8, 13, 28, 30, 70
inheritance, 6
inhibition, 26, 28, 41, 42, 71
inhibitor, 26, 31, 50
initiation, 3, 17, 18, 19, 28, 64
insertion, 5, 12
instability, 3, 26, 30, 33, 35, 64
insulinoma, 51
integration, 1, 5, 6, 24
integrity, 30
interaction, 3, 6, 12, 24, 35, 43
interactions, ix, 44
interference, 19
introns, 21
ionizing radiation, 6, 9, 10, 18
iron, 20
irradiation, 34, 57
ischemia, 19, 20
isozymes, 49

J

Japan, 7

K

KAI1, 44
kidney, 10, 15, 32, 49, 50
kidney tumors, 32
killing, 19
kinase activity, 27, 53

L

labeling, 51
larynx, 15, 16
latency, 9
lesions, 38
leucine, 28
leukemia, 2, 5, 16, 21, 24, 25, 27, 28, 32, 49, 50, 51
ligand, 23, 25, 27, 68
likelihood, 38
limitation, x
links, 7, 33
lipid peroxidation, 20
liver, 15, 16, 17, 25, 29, 35, 38, 43, 44, 48, 49, 50, 51, 52
liver cancer, 38, 44, 48, 49, 50, 52
localization, 48
locus, 7, 26, 29
long distance, 25
lumen, xi
lung cancer, 29, 35, 43, 50, 53
lymph, 43
lymph node, 43
lymphoma, 2, 16, 22, 24, 25, 32, 49, 50, 51

M

machinery, 7, 33
macromolecules, 12, 53
malignancy, 7, 30
malignant tumors, 1, 4, 13, 19, 47
mammography, 38
management, 64, 68, 69, 72
manufacturing, 16
marriage, 59
mastectomy, 38
matrix, ix, 41, 44, 51, 69
matrix metalloproteinase, 69
maturation, xi
meat, 15
melanoma, 10, 24, 50

mesothelioma, 10
metabolic pathways, 19
metabolism, x, xi, 1, 20, 30, 31, 34, 37, 53, 55, 57, 61
metabolites, 19, 53
metastasis, 3, 17, 41, 43, 44, 48, 58, 61, 67, 68
methylation, 44
mice, 13, 42
micronutrients, 55, 70
migration, 43
mitochondria, 8, 30, 63
mitogen, 27, 35, 44
mitosis, 34, 42
mitotic index, 51
MMP, 69
mobility, 41
molecular biology, 47
molecular weight, 20
molecules, 1, 31, 58
monoclonal antibody, 58
motion, 20
mRNA, 26, 29, 33, 68
mucin, 49, 70
mucosa, xi
multicellular organisms, 1
multiple myeloma, 49, 51
multipotent, ix
mutagen, 9
mutagenesis, 18
mutant, 7, 22, 23, 24, 30, 31, 32, 33, 34, 35, 38, 49, 63
mutation, x, 3, 6, 7, 10, 11, 12, 17, 19, 20, 23, 26, 27, 29, 30, 32, 33, 34, 35, 36, 38, 45, 47, 50, 63

N

necrosis, 58
neoplasm, 1
network, 43
neuroblastoma, 23, 24, 26, 29, 30, 49, 50, 51
neurons, x

Index

neutrons, 10
nicotine, 37
nitrogen, 11, 12
nitrosamines, 53
nodes, 43
non-smokers, 52
non-steroidal anti-inflammatory drugs, 38
nuclei, x
nucleic acid, 20
nucleotides, 30, 58
nucleus, 21, 33, 34, 41
nutrients, 9

O

obligate, 28
observations, 71
oil, 13
oncogenes, 3, 5, 6, 15, 20, 21, 22, 23, 24, 30, 34, 36, 41, 44, 49, 61
organ, 43
organism, ix, 1, 41
osteogenic sarcoma, 32
ovarian cancer, 7, 31, 32
ovaries, 50
oxidation, 12, 20
oxidative stress, 9, 18, 19
oxygen, 8, 10, 12, 18, 19, 58

P

p53, xi, 6, 11, 20, 25, 29, 32, 34, 35, 37, 45, 49, 58, 63, 65, 66
palliative, 57
PAN, 49
pancreas, 15, 24, 27, 48, 49, 50
pancreatic cancer, 27, 44, 63, 64, 71
parasitic infection, 9
particles, 10
pathogenesis, 45
pathways, 27, 28, 35, 59, 67
PCR, 68

penetrance, 20, 37
perfusion, 59
peritoneum, 16, 44
pharmaceuticals, 65
phenotype, 21, 42, 43, 45
pheochromocytoma, 23, 33, 51
phosphorylation, 8, 23, 35
plants, 1
plasma, 28
plasma membrane, 28
plasmid, 26
plasminogen, 50
pleura, 16
plexus, 43
point mutation, 5, 12, 23, 24, 26, 35
polarity, 41
pollution, 61
polycyclic aromatic hydrocarbon, 53
polymorphism, 26, 33, 37
polymorphisms, 38, 66
polypeptide, 51
polyphenols, 55
polyploidy, 30
polyps, 31
poor, 8, 23, 59, 63, 69
population, 37, 38, 48, 52
porphyrins, 10
post-transcriptional regulation, ix
power, 10
prediction, 48
preference, 20
pregnancy, 38
premature death, 6
pressure, 19
prevention, 38, 61, 69, 70
preventive approach, 55
probability, 35
proband, 38
production, 20, 26, 47, 50, 51, 57
progesterone, 50, 52
prognosis, xi, 2, 8, 23, 26, 48, 50, 63
program, 33

proliferation, x, xi, 1, 5, 11, 13, 20, 21, 25, 27, 28, 30, 36, 41, 43, 51, 53, 58, 71
promoter, 5, 11, 13, 25, 30, 33, 35
prophylactic, 38, 59
prostate, 1, 13, 15, 23, 33, 44, 48, 49, 50, 67
prostate cancer, 13, 23, 33, 44, 48, 49, 50, 67
protease inhibitors, 55, 58
protein kinase C, 13
protein kinases, 22
protein-protein interactions, 5
proteins, ix, 6, 8, 12, 21, 22, 23, 24, 25, 26, 27, 28, 36, 42, 47, 59, 66, 67
proteomics, 63
proto-oncogene, 5, 6, 7, 21, 22, 24, 25, 26, 27, 30, 35, 41, 45
prototype, 11
puberty, 13
purines, 12
pyrimidine, 10, 34, 42

R

radiation, 6, 9, 15, 16, 17, 18, 58, 71
radiation therapy, 71
radical formation, 20
radio, 43
radiotherapy, 57
radon, 9
range, 10, 19
reactants, 51
reactive oxygen, 20, 30, 34
reading, 34
receptors, 25, 26, 27, 32, 33, 34, 37, 41, 42, 50, 52, 58
recognition, 6
recombinant DNA, 47
recombination, 3, 7, 25, 32
rectum, 43
recurrence, 43, 48, 71
regulation, 35, 42
relevance, 69
repair, 3, 6, 7, 8, 32, 37, 38, 45

replication, 3, 7, 34
repression, 27, 28
resection, 48, 59
residual disease, 43, 48
residues, 35
resistance, 8, 9, 27, 34, 59
responsiveness, 27
retinoblastoma, 32, 35, 37, 49, 66
retrovirus, 5, 24
retroviruses, 5, 29
returns, 29
reverse transcriptase, 5
risk, 6, 36, 37, 38, 48, 59
risk factors, 37
RNA, 5, 12, 21, 28, 30, 42, 63

S

satellite, 26
scatter, 22
schistosomiasis, 38
scrotum, 10
secrete, 1
secretion, 41, 42
segregation, 7
selenium, 35, 55
sensitivity, 6, 38, 52
serine, 27, 28, 31, 49, 50
serum, 48, 53
serum albumin, 53
sialic acid, 49, 51
signal transduction, 5, 21, 22, 31, 32, 67
signaling pathway, 44, 71
signals, 26, 27
sinuses, 16
sister chromatid exchange, 53
skin, 1, 6, 13, 16, 22, 30, 32, 38, 49
skin cancer, 30, 32, 38
smoke, 15
smokers, 52
smoking, 15, 38, 58
smoking cessation, 38

soft tissue sarcomas, 68
solid tumors, 2, 19, 32
somatic cell, 36
somatic mutations, 45
specialization, ix
species, 18, 20, 30, 34
specificity, 52
spindle, 57
stability, x, 18
stabilizers, 20
stages, 17, 18, 26
stem cells, ix, x, xi
stomach, 6, 49, 55
strategies, 26
stress, 19, 31, 67
stroma, 67
supply, x
suppression, 34
surgical resection, 59
surveillance, 52
survival, 1, 25, 71
susceptibility, 7, 9, 32, 38, 48, 53
syndrome, 6, 7, 30, 31, 32, 35, 37
synthesis, 30, 42, 55, 57, 58

T

T cell, 49
targets, 67
telangiectasia, 32
telomere, 34
telomere shortening, 34
tension, 8
testicle, 50
TGF, 27, 32, 50
therapy, 38, 48, 50, 58, 68, 71
threonine, 23, 27, 28
threshold, 3, 17, 45, 61
thymine, 18
thyroid, 16, 23, 24, 31, 49, 50
thyroid cancer, 23, 31, 50
tissue, ix, x, 1, 4, 19, 20, 47, 48, 53, 61

TNF, 37
tobacco, 10, 11, 12, 38, 53, 69
tobacco smoke, 11, 12
TPA, 13, 51, 52
transcription, ix, 7, 21, 22, 29, 31, 32, 33, 34, 36, 68
transcription factors, ix, 21, 22, 32, 36
transcripts, 68
transfection, 21, 44
transformation, x, xi, 3, 6, 13, 17, 20, 21, 25, 28, 29, 30, 36, 41, 45, 48, 53, 61, 67
transforming growth factor, 70
transition, 30, 34
transitions, 12
translation, 29
translocation, 5, 24, 25, 29
trophoblastic tumor, 50
tumor, xi, 1, 3, 5, 6, 7, 9, 11, 13, 17, 18, 19, 20, 25, 27, 30, 31, 32, 33, 34, 36, 37, 42, 43, 44, 45, 47, 48, 49, 50, 51, 52, 53, 55, 57, 61, 63, 64, 66, 69, 70
tumor cells, 3, 27, 33, 34, 44, 47, 57, 70
tumorigenesis, 3, 68
tumors, 1, 4, 20, 22, 26, 27, 29, 32, 33, 35, 43, 49, 50, 67, 69
turnover, 51
twins, 38
tyrosine, 22, 23, 26, 27, 31, 53

U

ulcerative colitis, 38
urine, 53
uterus, 6, 49
UV, 6, 9, 10, 15, 34, 38
UV light, 6
UV radiation, 9, 10, 34

V

vagina, 15
values, 52

vascular endothelial growth factor (VEGF), 31
vegetables, 55
villus, xi
virology, 66
viruses, 5, 9, 15, 21, 24, 30, 63, 64
vitamin C, 55
vitamin D, x, 44
vitamin E, 55

W

warts, 22
women, 38, 48, 59

X

xenografts, 64

Z

zygote, ix